Rice in a Variable Climate

Yash P. Abrol
Sulochana Gadgil

2024

Associated Publishing Company®

A Division of

Astral International Pvt. Ltd.

New Delhi – 110 002

First Published, 2019

Reprinted, 2024

ISBN: 978-81-86580-34-9 (HB)

Published by : **Associated Publishing Company®**
A Division of
Astral International Pvt. Ltd.
– ISO 9001:2015 Certified Company –
4736/23, Ansari Road, Darya Ganj
New Delhi-110 002
Ph. 011-43549197, 23278134
E-mail: info@astralint.com
Website: www.astralint.com

Printed at : Replika Press Pvt. Ltd.

RICE—in a Variable Climate

EDITORS

Yash P Abrol
Division of Plant Physiology
Indian Agricultural Research Institute
New Delhi 110012

Sulochana Gadgil
Centre for Atmospheric and Oceanic Sciences
Indian Institute of Science
Bangalore 560012

Preface

Rice is the primary food of the Asians and over 85 per cent of the rice is produced in this continent. India ranks second in both production and area in the World. It contributes over 40 per cent to the annual food grain production of the country.

The crop is grown over a wide range of climatic conditions between 45°S and 40°N and over land from sea level up to 3000 m and under extreme moisture regimes ranging from low rainfall (<400 mm) drought prone upland to high rainfall (>3000 mm) deep water area of eastern India. The area for rice cultivation is 43 mha of which only 19 mha is irrigated. While the average yield under irrigation, particularly in the regions with ample sunshine, are high, yields under rainfall ecologies are poor and show a high degree of instability and depend heavily on the vagaries of the monsoon. Thus the major challenge before the agricultural and atmospheric scientists how to increase production of rainfed areas. This is particularly so, as the rate of increase in production of the high input regions is decreasing indicating green revolution 'fatigue' and questions are being raised about sustainability of the system. Furthermore, the demand for food production is on the way up.

Against the above background and as a follow up to the earlier workshop (See *Climate Variability and Agriculture*, YP Abrol, S Gadgil and GB Pant), which dealt with the overall linkages between climate variability/change and agriculture, a second workshop was held wherein rice experts, climatologists and crop modelers participated to assess the present status of knowledge pertaining to links between rice cultivation under diverse ecologies and climatic variability, and identify constraints. Because of implication of paddy cultivation, particularly under flooded conditions, in emission of methane, 'a green house gas', experts working in this area also participated.

This volume comprising of eleven chapters is based on the presentations during the workshop. The topics range from 'origins of rice', analysis of linkages between climate variability/change and rice productivity/production, use of various crop models, incidence of pest and fungal diseases to methane emissions. The last chapter entitled 'Rice in 2000 A.D' by M.S. Swaminathan published in the Golden Jubilee Volume of the Indian National Science Academy is reproduced. In some of the chapters, the contributions, by individual author(s) on related fields have been synthesized as to avoid overlapping and present a coherent picture.

The editors/organizers wish to thank the World Climate Research Program, the Jawaharlal Nehru Centre for Advanced Scientific Research, Centre for Atmospheric Sciences and Centre for Ecological Sciences of Indian Institute of Science, the Indian Council of Agricultural Research, the International Geosphere and Biosphere Program and the Global Change System for Analysis Research and Training (START) for their financial support.

Yash P. Abrol
Sulochana Gadgil

Contents

Rice—in a Variable Climate, 1999, pp. 1-10
Eds. Y.P. Abrol and Sulochana Gadgil

1. Origins of Rices

Rice and wheat are the most important food crops of human kind. Of these, rice should be called the *staff* of life, because rice constitutes the staple food of more people in the World than wheat. Worldwide, rice is grown in about 11 per cent of the total available land (1,350 million ha; wheat: 16 per cent), but unlike wheat, 95 per cent of the world's rice is grown in less developed countries. Asia accounts for 90 per cent of the area and 92 per cent of the production. China and India rank one and two in both area and production. Together, they account for about half of both the area and production. Outside Asia, rice is an important food crop in Guyana, Guinea, Liberia, Sierra Leone and Madagascar (IRRI 1995).

Taxonomically, two distinct species are cultivated in the World—the Asian cultivated rice, which is the dominant rice of the World, *Oryza sativa* L. and the West African cultivated rice, *Oryza glaberrima* Steud. The latter species is, however, being replaced by *O.sativa* because of its wider adaptability and higher yields.

Both the cultivated rices belong to the genus *Oryza* L., tribe Oryzeae, subfamily Bambusoideae, family Gramineae (Poaceae). The genus *Oryza* is a small one with about 20-22 species. The first classification of the genus was done by Prodoehl (1922), who included 17 species, and the latest has been by Vaughan (1989), who included 23 species.

With the availability of biosystematic information, rice taxonomists have been of late grouping the species of the genus into species complexes. They are *O.sativa* complex, *O.officinalis* complex, *O.meyeriana* complex, *O.ridleyi* complex, and two unattached species.

Both the cultivated rices are diploids with 2n = 24 chromosomes. About two-thirds of the *Oryza* species are diploids (2n: 24) and the remaining tetraploids (2n: 48), except for one, *O.schlechteri*, whose chromosome number has not been determined yet.

Vaughan (1989) has eight species including the two cultivated species in the *O.sativa* complex. All of them have the basic A genome (Table 1).

Considerations on crop plant origins involve covering four major aspects-place of origin, time of origin, species from which originated and mode of evolution of cultivated species from its ancestral forms. Evidence for these come from several sources—cytogenetical, biosystematical, archaeological, geological, philological, and accounts of ancient travels and human migrations. It is only rarely that any single source would give clinching evidence on any of these aspects.

Table 1. Species of *sativa* complex*

Species name		Genome composition
i.	*O.sativa*	A
ii.	*O.nivara*	A
iii.	*O.rufipogon*	A
iv.	*O.glaberrima*	A^gA^g
v.	*O.barthii*	A^gA^g
vi.	*O.longistaminata*	A^lA^l
vii.	*O.glumaepatula*	$A^{gp}A^{gp}$
viii.	*O.meridionalis*	A^mA^m?

* From Vaughan (1989).

We shall now consider the two rices separately.

1.1. *O.sativa* L., The Asian Cultivated Rice

No other cultivated plant shows such tremendous range of variation and adaptation as *O.sativa*. Notwithstanding its high adaptability, it grows best in tropical wetland conditions having plentiful supplies of water.

In rice, archaeological evidences have not been able to give any definite leads on the origin of rice. Several reasons have been ascribed. The hot and humid climates that prevail in most of the rice growing areas are not conducive for long term preservation of biological specimens. *Secondly*, the morphological differences between the cultivated rice and its putative ancestors are very slight, and the archaeological samples are not likely to be adequate to determine with any certainty their correct species identity. *Thirdly*, many of the excavated samples have not been correctly dated, particularly from China. *Fourthly*, the nationalistic feelings and over-enthusiasm for establishing a name seem to have clouded some workers' judgement and interpretation of certain findings.

1.1.1. Place and Time of Origin

The two 'fathers' of crop plant origins held differing views on the origin of rice. De Candolle (1882), a famous botanist, using also evidences from other sources presumed that rice culture in China was older than that of India. On the other hand, Vavilov (1926, 1951), the pioneer botanist in biodiversity conservation, who used also evidences on plant biodiversity and field observations, was of the view that rice originated in India.

The oldest evidence from published archaeological records show that rices have come from India, China and Thailand (Table 2). There is lot of data from various sites in India and China belonging to later periods (cf., Oka 1988). The broad inference that could be arrived at is that since cultivated and wild rices occur widely and sympatrically and since their spikelet morphologies are broadly similar, it would be difficult to identify any specific site as the place of origin of rice.

Table 2. Oldest archaeological evidences

Authors	Age	Site .	Nature of material
Sharma & Manda (1980)	6570 ± 210 to 4530 ± 185 BC (carbon-dated)	Maharaga, UP, near Belen river, India	Neollithic site: earthenware having rice grains, length/ breadth ratio of grains 2.15.
Ghosh (1961)	2300 BC (carbon-dated)	Lothal, Gujarat, India	Remnants of husk, impressions of paddy on clay lumps
Wang (1986)	5060 BC	Zejiang, Yuyao County, China	Rice grains mostly carbonized
Solheim (1972)*	About 6000 BC	Non-Nok-Tha, northeast Thailand	Cord-marked potteries
Gorman (1977)*	do	Spirit-Cave, northwest Thailand	Dryland paddy culture, slate knives and other tools used in rice harvest

*These two are considered as part of Hoabinhian culture.

1.1.2. The Ancestral Species

There is general agreement that the Asian cultivated rice. *O.sativa* L.—has originated from the perennial floating wild rices of Asia. This taxon is referred to in the literature by several names—*O.balunga*, *O.perennis*, *O.perennis* var. *fatua*, *O.perennis* var. *balunga*, *O.fatua*, *O.sativa* f. *aquatica*, *O.formosana*, *O.aquatica*, *O.sativa* ssp. *rufipogon* and so on—, but Clayton (1968) has convincingly established that its correct nomenclature is *O.rufipogon* Griff. The species occurs extensively throughout south, southeast and east Asia, and north Australia. It is a perennial floating plant found in permanent water bodies such as tanks, samples, canals, deep ditches, and so on. In the last four decades, because of extensive urbanization and human activities, most of the natural stands have either disappeared or badly fragmented. Interestingly, until about two decades back, when food supplies were less adequate, it was not uncommon to see people harvesting grains from areas having large stands of *O.rufipogon* by swaying a basket over the flowering panicles (Nayar 1958; Oka 1988). And more interestingly, a similar practice has been in vogue also in tropical West Africa (Oka 1988) of collecting wild rices from areas where natural wild rice stands occur extensively.

The grains of wild rices, *O.rufipogon* are essentially similar to those of cultivated rices; only that they are more slender, red grained (rices grown in Malabar coast and south eastern Indian costal regions are also red grained by consumer preference), awned (rices grown in deep water and saline areas also possess awns), and are shed easily—which is nature's way to facilitate easy dispersal (Table 3). Wild rices occur sympatrically with cultivated rice. While cultivated rices are predominantly self-fertilized, wild rices are largely cross-fertilized; pollination taking place by wind. Wherever *O.sativa* is cultivated in close proximity to the habitats of *O.rufipogon*, natural crossings between the two species take place. Hybrids between the two species i.e. the cultivated *O.sativa* and the wild *O.rufipogon*, are common throughout south, southeast and east Asia, and to a less extent in north Australia. They occur in cultivated rice fields and adjacent water channels vacant and fallow fields, ditches, etc. They are generally annual in habit and show characters that are intermediate between the two species. (But, it should be kept in mind that even cultivated Asian rices are not strictly annual in habit, and, so are these wild rices also). They have been a nightmare to the taxonomists. They have been given either specific, subspecific or varietal epithets as *spontanea*, *fatua*, *balunga*, *nivara*, *meridionalis*, *glumaepatula*, *paraguayensis*, *cubensis* and so on (of either of the two putative *parental* species), or simply left without assigning any name with various remarks as hybrid *O. sativa*, wild rice, etc.

Experimental studies have shown that most, if not all, plants of *spontaneas* (this epithet will be used in this paper for the material that is morphologically intermediate between *O.sativa* and *O.rufipogon*) are both heterogeneous and heterozygous in nature. Further, they look indistinguishable from the segregating populations obtained from hybrids between *O.sativa* and *O.rufipogon*. New species that have been described recently as *O.nivara*, *O.meridionalis* and *O.glumaepatula* look deceptively similar to these *hybrid* and variable populations that has been designated here as *spontaneas*.

Table 3. Characteristics of *O.sativa* and *O.rufipogon*

Character	*O.sativa*	*O.rufipogon*
Habit	Typically annual, erect, tillers from basal nodes, no extra-vaginal branches	Perennial, typically floating and decumbent-extravaginal branches and roots from all nodes present
Panicle	Typically compact, heavier with more spikelets, and awnless	Open, fewer and more slender spikelets
Anthers	± 2.0 mm long	± 3.0 long
Breeding behaviour	Almost completely self-fertilized	Predominantly cross-fertilized
Spikelets	Generally awnless, resistant to shedding, golden brown in colour, grains of most varieties white, 4.0-8.5 mm long, 2.0-4.0 mm wide	Generally awned, easily shed, mostly black in colour, grains dark red, 7.0-9.3 mm long, 1.9-2.6 mm wide

1.1.3. Mode of Evolution

Some authors have suggested that the evolution of *O.sativa*, from the perennial *O.rufipogon* would have been mediated through an annual wild species. Experimental, including transplant studies, have shown that this does not have to be so. Almost every variety of the cultivated rice, can be made to regenerate indefinitely for any number of years, if favourable cultural conditions like water and nutrients are given to the stubbles. Oka (1988), who has worked on wild and cultivated rices all his lifetime, has stated that "*O.sativa* is potentially a perennial plant" This characteristic of cultivated rice, *O.sativa* is quite unlike that of almost all other cultivated cereals including even the African cultivated rice *O.glaberrima*. Hence, there is no need to postulate the evolution of *O.sativa* from the perennial *O.rufipogon* through an intermediate annual species (Nayar 1958; Sampath and Govindaswami 1958).

Let us now consider the mode of evolution of the cultivated *O.sativa* from its progenitor species. Though the gross morphology of the two species are broadly similar, if we have to name one character that essentially separates the cultivated species from *O.rufipogon*, it is the easy shedding nature of spikelets of the wild species and the relative resistance to shedding of spikelets of the cultivated species. Several genetic studies have shown that this character is controlled by a pair of complimentary genes (Nayar 1958, 1973). This could have been obtained either by recombination or by mutation at one of the two involved *loci*. With such an important primary gain, incorporation of other important economic characters in the cultivated forms would have been brought about progressively. The present author had proposed (Nayar 1973) that the evolution of *O.sativa* from *O.perennis* can be seen to be an instance of disruptive selection, and that it has had a 'diffuse origin', since the essential process of differentiation of *O.sativa* from *O.rufipogon* could have been brought about by a simple gene mutation initially, and since *O.rufipogon* occurs over most major areas of rice cultivation in south, southeast and east Asia. This means that the process of evolution of *O.sativa* could have happened many times both in space and time. This point of view has been reiterated by others subsequently (Chang 1995).

1.2. *O.glaberrima* Steud, The African Cultivated Rice

This is the traditional cultivated rice of Africa, specifically tropical west Africa, extending from Senegal to north Cameroon and Chad. Unlike the predominant position of rice in the food economy of Asia, rice is only of minor relevance in that of tropical Africa, providing less than 8 per cent of the calorific intake. The population, in addition to being primarily non-vegetarian, use other cereals, and root and tuber crops as their main sources of carbohydrate—sorghum, millets (pearl and finger millets) fonio (*Digitaria exilis*), cassava and yams. Further, unlike 'rice Asia,' tropical West Africa is only sparsely populated, and there are vast stretches of savannah grasslands that have not been subject to much of human interference.

Both morphologically and genetically, *O.glaberrima* is distinguished from *O.sativa* (Table 4). *O.glaberrima* is generally much less hairy, the panicles have no or only few secondary branches, ligule is short and is strictly annual as compared to the Asian rice. Interestingly, *O.sativa* has almost completely replaced *O.glaberrima* because of its wider adaptability and generally higher yields.

The hybrids between the two species are sterile, though F_1 hybrids are pollen sterile, but show partial female sterility. The sterility is assumed to be genetically controlled (Oka 1988). Oka and coworkers, who have made several field collection trips in this region; have observed that it is common to see mixed stands of the two cultivated species in cultivators' fields, and also mixed stands of African rice and the wild rices (cf., Oka 1988).

Table 4. Comparision of the two cultivated species and their near relatives

Characters	*O.sativa*	*O.glaberrima*	*O.barthii*	*O.longistaminata*
Habit	Annual, but can be propagated perennially	Strictly annual.	Strictly annual	Strongly perennial.
Panicle	Generally compact & heavy, with secondary and tertiary branches	Nearly compact, less heavy, secondary branches only sometimes present	Open, less heavy, secondary and tertiary branches present	Open, not heavy, secondary branches present
Spikelets	Generally awnless, resistant to shedding, 4.0-8.5 × 2.0-4.0 mm	Generally awnless, less resistant to shedding, 7.4-9.0 × 2.9-3.6 mm	Awned, easily shedding 7.8-1.0 × 2.8-3.4 mm	Awned, easily shed 8.0-8.5 × 2.4-2.6 mm
Anther length	2.5-2.6 mm	1.5-.8 mm	1.8-2.0 m	>4.0 mm, almost filling the spikelets

Source: Compiled from several published sources.

1.2.1. Wild Rice Species in Tropical West Africa

In tropical West Africa, two wild *Oryza* species occur and both are widely occurring. They are the strongly perennial and rhizomatous *O.longistaminate* Cheval. *et* Roehr. and the annual *O.barthii* Cheval.

Until Clayton (1968) correctly determined this nomenclature, the perennial species has been known in literature by various names as *O.barthii* (most widely used name), *O.perennis*, *O.perennis* ssp. *barthii*, *O.dewildemanii*, *O.perennis* ssp. *madagascariensis*, etc., and the annual wild species as *O.breviligulata* (most widely used name), *O.mezii*, *O.stapfii*, *O.glaberrima* spp. *barthii*, and so on.

Both these wild species and the cultivated *O.glaberrima* are diploids (2n: 24), they are intercrossable, but strong sterility barriers exist among them. The sterility has been attributed to genetical factors (Oka 1988). Inspite of this, natural hybrids between *O.glaberrima* and *O.barthii* are frequent, and some workers have proposed that the taxon known in literature as *O.stapfii* is derived from this cross. Here also, as with the Asiatic wild rices, wherever large stands of African wild rices occur, people harvest them commonly for the grains.

1.2.2. Time and Place of Origin

Archaeologial evidence from Sub-Saharan region are weak or nonexistent. This may be primarily because there might not have been any significant human settlements in these regions in 'archaeological times'. It is sparsely populated even now. The French botanist, Porteres, who had done considerable field studies in this region in the 40's and 50's, had postulated that African rice originated in the Central delta region of Niger river at about 1500 BC. The area is savannah grassland with marshes and small rivers, and it is conducive basically to wetland rice culture. It is also well known that several points along the Niger including Lake Chad, were major staging centres of ancient caravan routes between North Africa and tropical West Africa.

1.2.3. The Ancestral Species

Most workers have assumed that the annual wild rice, *O.barthii* (long known in literature as *O.breviligulata*) is the progenitor of *O.glaberrima* (cf., Oka 1988; Chang 1995).

Some botanists, however, who have had extensive knowledge of African agriculture including rice culture (eg. Chevalier, Porteres; cf. Nayar, 1973 for reference) have proposed that two other species, *O.longistaminata* and *O.sativa*, have also contributed in some measure to the evolution of *O.glaberrima*. There are also at the same time, some strong contra-indications on *O.barthii* being the progenitor species of the African cultivated rice *O.glaberrima*. For instance, *O.barthii* has a far wider range of distribution and possesses much greater variability than *O.glaberrima* (Second 1986, Oka 1988). The general observation is that the cultivated species show greater genetic and morphological variation and far wider range of distribution cultivation than their respective ancestral species.

Further, data gathered by various workers show that the spikelet size, and often, even the total yield/panicle are greater in *O.barthii* than in *O.glaberrima* (cf., Nayar 1973). It is inconceivable that man in his quest for improving yield and other economic attributes would select for smaller grain size and yield/panicle, if it is assumed that evolution of *O.glaberrima* has proceeded from *O.barthii*.

It has, therefore, been postulated (Nayar 1973) on the basis of considerable convincing, albeit indirect, evidence that *O.glaberrima* has evolved from the Asian cultivated rice, *O.sativa*.

In recent times, the Asian rice, *O.sativa* is believed to have been introduced into tropical West Africa in the 15th century by the Portuguese. Much earlier to this, after the Asian rice *O.sativa* was first introduced into Africa to Egypt, it was taken westwards to North Africa. From here, almost from very early times, there used to be regular movement of people and merchandise between North Africa and tropical West Africa through several caravan routes (Murdoch 1959). It could be assumed logically that *O.sativa* too would have been brought to tropical West Africa from North Africa during these movements (cf. Nayar 1973). Taking into account the narrow genetic base of *O.glaberrima*, its restricted area of cultivation and its limited adaptability, it can be further deduced that this species is of only comparatively recent origin. Incidentally, the 3-4 characteristics that distinghuish *O.glaberrima* from *O.sativa* (Table 4) have all been reported from the Indian cultivated rices collected from northeast India, Jeypore region and elsewhere.

1.2.4. Mode of Evolution

It has been further postulated that *O.glaberrima* has the nature of a species vis-a-vis *O.sativa* (Nayar 1973). Sibling species are "sympatric populations that are morphologically similar or identical, but are reproductively isolated (Mayr (1963). Numerous examples of sibling species have been given in both plant and animal kingdom by such well known workers as Dobzhansky, Grant and others. (cf. Nayar 1973; Grant 1971).

Regarding the mode of differentiation, it has been suggested that *O.glaberrima* has evolved by a process designated as quantum speciation. It is defined as the budding off of a new daughter species from a semi-isolated peripheral population of an ancestral species" (Grant 1971).

All the cytological, genetical and botanical studies point to close relationship among all these four taxa—*glaberrima, sativa, barthii* and *longistaminata*—and also with the form referred to in literature as *O.stapfii.*

A corollary to these is the possible hybrid origin of *O.barthii* from *O.sativa-O.glaberrima* hybrids. Forms resembling and not morphologically distinguishable from the naturally occurring *O.barthii* can be routinely obtained in the experimentally crossed progenies of *O.sativa* and *O.glaberrima.* Broadly, then *O.barthii's*—and the taxa known as *O.sapfii's* position is analogous to that the *spontaneas*, sometimes known in literature as *O.nivara.*

1.3. References

CHANG, T.T. 1976. *The origin, evolution, cultivation and dissemination of Asian and African rices. Euphytica.* **25**: 425-441.

CHANG, T.T. 1995. Rice p. 147-155. In: *Evolution of Crop Plants* (eds. J Smart and NW Simsonds). Longman, Burnt Mill, UK. pp. 531.

CLAYTON, W.D. 1968. *Studies in the Gramineae*-17. Kew Bull. **21**: 485-488.

dE CANDOLLE, A. 1886. *Origin of cultivated Plants.* Reprint. Hafner, New York, 1967. pp. 468.

GHOSH, S.S. 1961. *Further records of rice from ancient India. Indian Forest.* **87**: 295-301.

GORMAN, C. 1977. *A priori models and Thai prehistory.* In: *Origins of Agriculture* (ed. CA Reed). *Mouton*, Paris. p. 321-355.

GRANT, V. 1971. *Plant Speciation.* Columbia Univ. Press, New York. pp. 435.

IRRI, 1995. *Rice Factsheet.* International Rice Res. Inst., Los Banos. pp. 6.

MAYR, E. 1963. *Animal Species and Evolution*. Harvard Univ. Press. Cambridge, M.A. pp. 797.

MURDOCH, G.P. 1959. *Africa*. McGraw Hill, New York. pp. 456.

NAYAR, N.M. 1958. *Studies on the Origin of Cultivated Rice*. Thesis. Indian Agric. Res. Inst., New Delhi. p. 109.

NAYAR, N.M. 1973. *Origin and cytogenetics of rice. Adv. Genet.* **17**: 153-292.

OKA, H.I. 1988. *Origin of Cultivated Rice*. Elsevier, Amsterdam. p. 254.

PRODOEHL, A. 1922. Oryzeae monographee describunhen. *Bol. Arch.* **1**: 221-224, 231-232.

SAMPATH, S. and GOVINDASWAMI, S. 1958. Wild rices of Orissa and their relationship to the cultivated varieties. *Rice Newsletter.* **6(3)**: 17-30.

SECOND, G.C. 1986. Isozymes and phylogenetic relationship in *Oryza*. *Rice Genetics.* **3**: 27-29.

SHARMA, G.R. and MANDA, D. 1980. *Excavation of Mahagara 1977-78*. Dept. of Ancient History and Archaeology. Univ. of Allahabad, Allahabad. pp. 278.

SOLHEIM, W.T. 1972. An earlier agricultural revolution. *Sci. Amer.* **226**: 34-41.

VAUGHAN, D.A. 1989. The genus *Oryza*, L. International Rice Res. Inst., Los Banos. pp. 21.

VAVILOV, N.I. 1951. Phytogeographic Basis of Plant Breeding. Vol. 13. Chronica Botanica, Waltham, M.A. (Includes his 1926 paper also). pp. 366.

WANG, T.T. 1986. *Discussion on origin and spread of agriculture in China.* (In Chinese) (from Oka 1988). *Agri. Archaeol.* **1986-2**: 25-32.

Rice—in a Variable Climate, 1999, pp. 11-23
Eds. Y.P. Abrol and Sulochana Gadgil

2. Rice Production in Variable Climate: Problems and Prospects

The strong links between agricultural productivity and climate variability over the Indian region leading to "the Indian economy being a gamble on the monsoon rains" have been recognised for most parts of this century. However, a major change in the perceived importance of climate/weather in agriculture occurred with the green revolution.

The rapid increase in the foodgrain production during the green revolution made it possible for the country to move from a food-deficit state to one which is by and large self-sufficient in food production, despite the marked increase in the population. The green revolution involved substantial gains in food production from well-endowed areas i.e. areas where irrigation already existed or could be developed relatively rapidly (Abrol 1994). These were areas with relatively few soil and climate associated constraints limiting production. The adoption of dwarf high yielding fertilizer responsive varieties (of rice and wheat, in particular) and a significant increase in the irrigation and fertilizer application led to a very large increase in the productivity of these regions and hence in the foodgrain production of the country in general (Fig. 1) and production of rice and wheat in particular (Fig. 2). Thus the rapid increase in food production during the green revolution occurred primarily in controlled environments and, not surprisingly, much attention was paid to weather/climate factors by the scientists involved.

The rate of growth of food production appears to have declined in the last few years. The limits of growth in food production from the high potential areas associated with the "fatigue of the green revolution" may well be in sight. At present, land area under irrigation is growing significantly more slowly than in the past two or three decades and in the future the rate of irrigated land going out of production may well exceed the rate of new land brought under irrigation (Ryan 1994). Thus to meet the increasing demands on food and feed in the future, the pressure on rainfed agricultural lands will increase. In contrast to the well endowed lands, progress has been

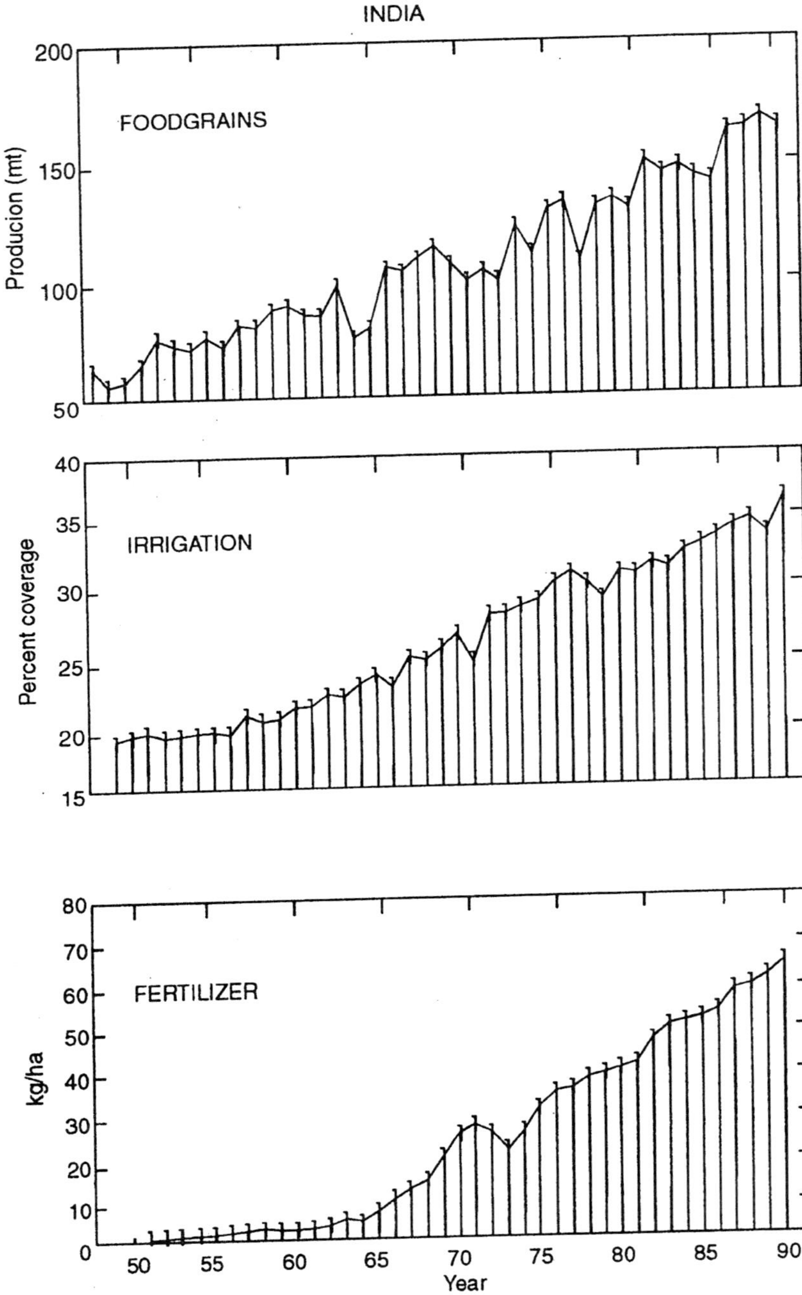

Fig. 1: Variation in Indian foodgrain production (million tons), irrigation (per cent coverage) and fertilizer (kg/ha) during 1950-91.

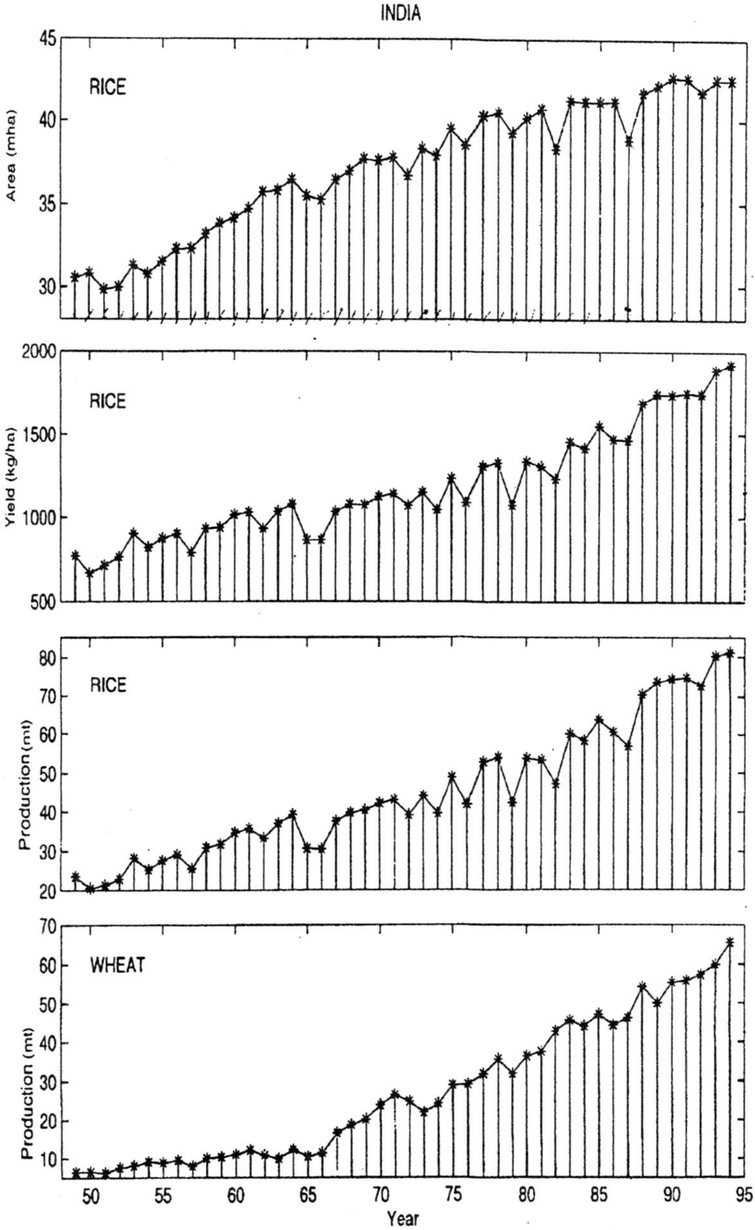

Fig. 2: Variation of area under cultivation for rice (mha), rice yield (kg/ha), rice production (million tons), wheat production (million tons) during 1950-1995.

very slow in the rainfed agricultural belt, which account for over 65 % of crop hectarage and almost half of the total crop production in the country (Abrol 1996).

For example, there has been hardly any increase in the production of pulses, which are the major source of protein, as they are grown primarily in rainfed regions. Hence, whereas the rate of growth in the production of the foodgrains has been adequate to ensure a reasonably stable per capita availability of food grains (wheat and rice) since the early fifties, the per capita availability of pulses has decreased by almost 50% (Fig. 3).

It is important to note that while the mean foodgrain production has increased since the sixties, the variability from year to year has not decreased. Superposed on the increasing trend due to the green revolution, are dips in the production of foodgrains and rice (e.g., 1965, 74, 79, 82, 87). Each such local minimum corresponds to a decrease of about 15-25% of the production peak in the next year. Rice production of various tropical countries such as China, Thailand, Bangladesh and Srilanka is also characterized by large inter-annual variation, (Deteng and Shaokai 1995, Surarerks 1995, Fagi and Las 1987, Yoshino 1993). The variation of the area under rice production for Thailand and Indonesia over the last three decades are shown in figures 4 and 5, respectively. The inter-annual variation in the rice production over Thailand is seen to be large, whereas that for the Indonesian rice production is rather small. The fatigue of the green revolution is apparent for the rice production in Thailand.

Climate variability, particularly rainfall variability, is the major factor influencing the agricultural productivity and sustainability in the tropics (Virmani 1994). The strong links between the vagaries of the monsoon and the variation of the total production of foodgrains and rice over the Indian region and the all-India summer monsoon rainfall with dips in production in years of poor monsoon are evident in Fig. 3. Note that there has been hardly any variability in the production of wheat, which is grown primarily under irrigated conditions in the last three decades. We expect the large variations in other parts of Asia such as Thailand to be also related to the variations in the rainfall. It is interesting that the inter-annual variations in rainfall do not seem to affect the rice production in Indonesia (Fig. 5). Even the EI Nino events of 1972, 82 and 87 (which were also associated with a poor monsoon over India and had large impact on the Indian rice and foodgrain production) do not appear to have had any impact on the rice production. This may be because the maritime continent is characterized by very heavy rainfall (Fig. 6) and even in years characterized by large deficits, the rainfall may still be adequate for rice production. However, over most of the other regions climate variability appears to have a significant impact on the rice production.

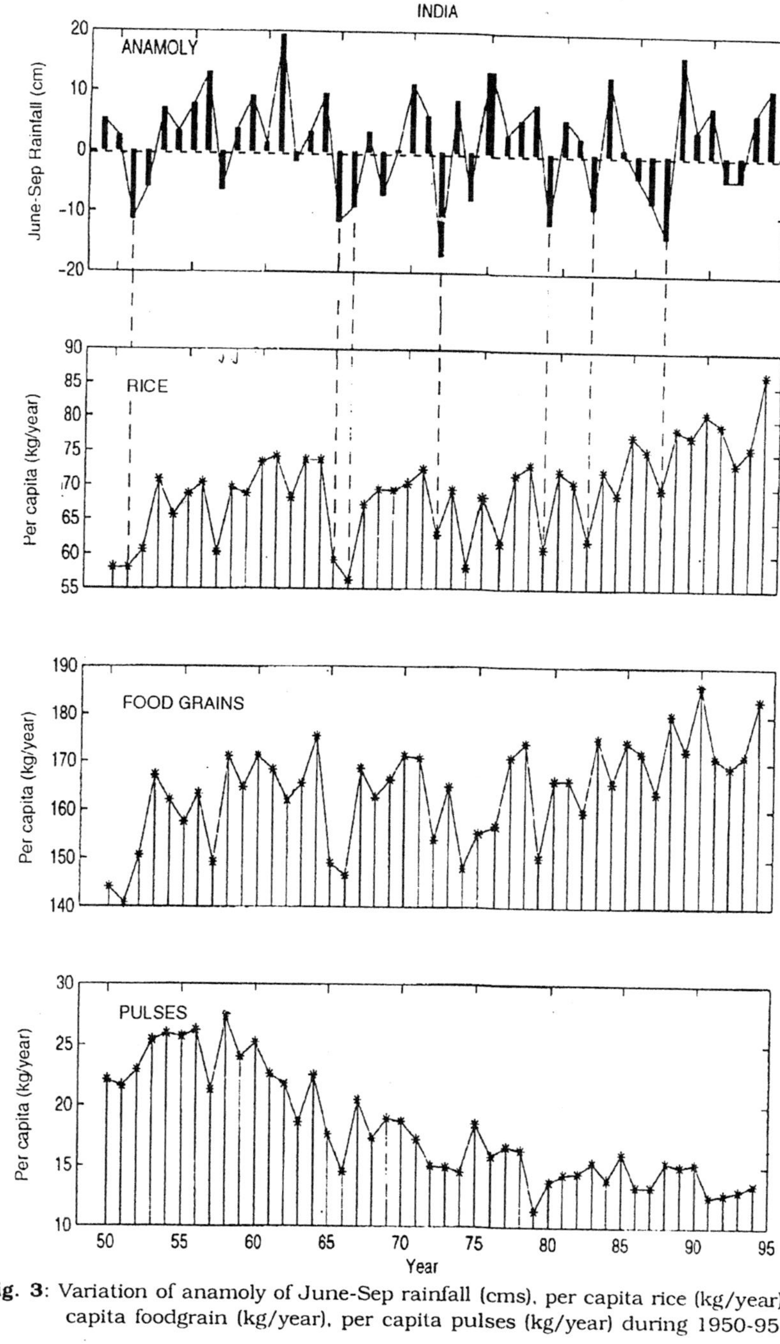

Fig. 3: Variation of anamoly of June-Sep rainfall (cms), per capita rice (kg/year), per capita foodgrain (kg/year), per capita pulses (kg/year) during 1950-95

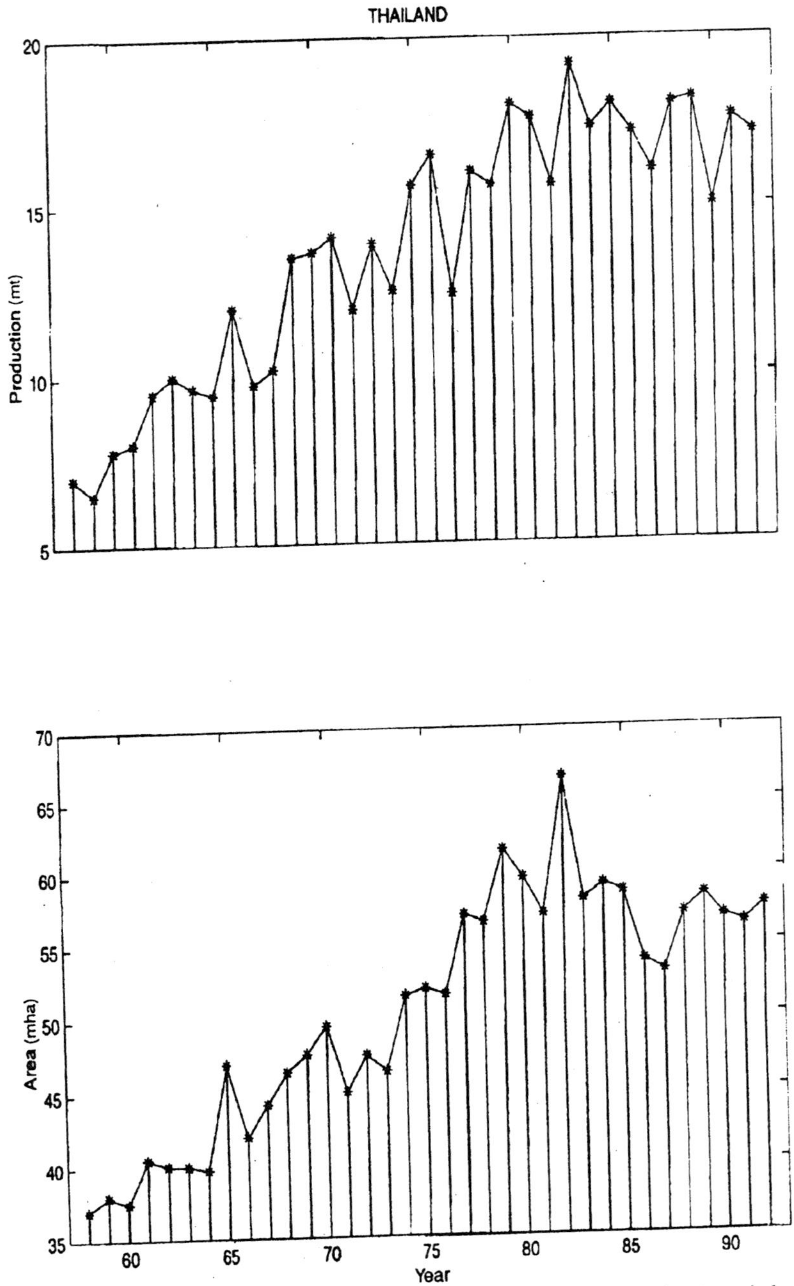

Fig. 4: Thailand rice production (million tons) and area (in million hectares) during 1958-1992.

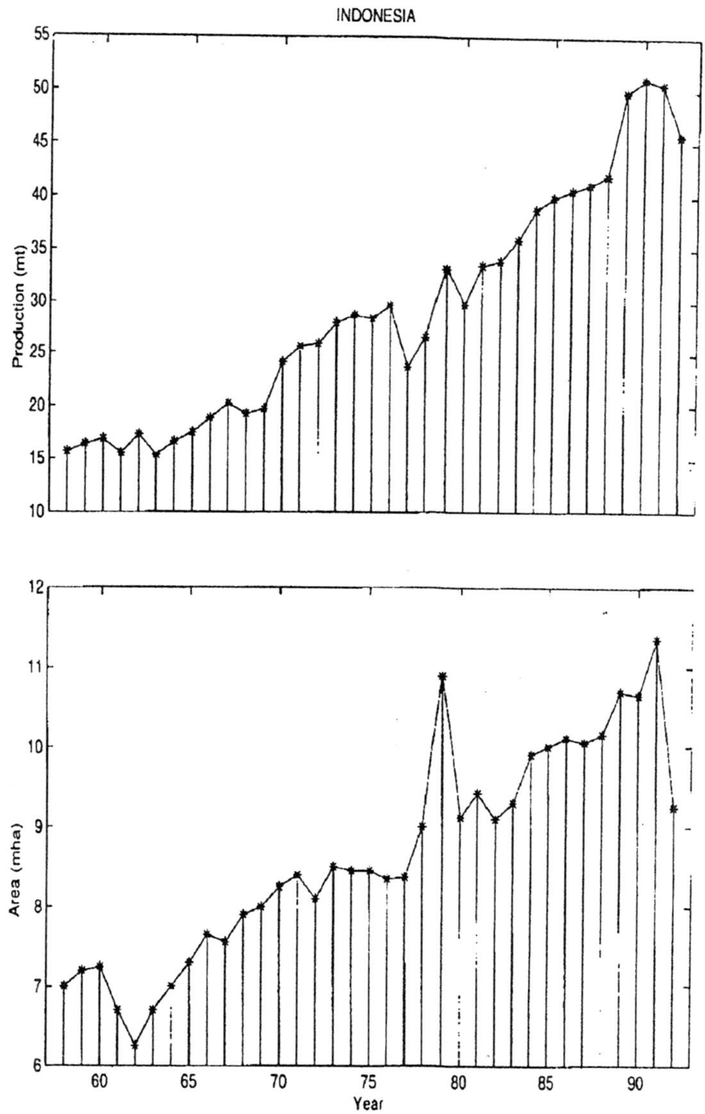

Fig. 5: Indonesian rice production (million tons) and area (million hectares) during 1958-1992.

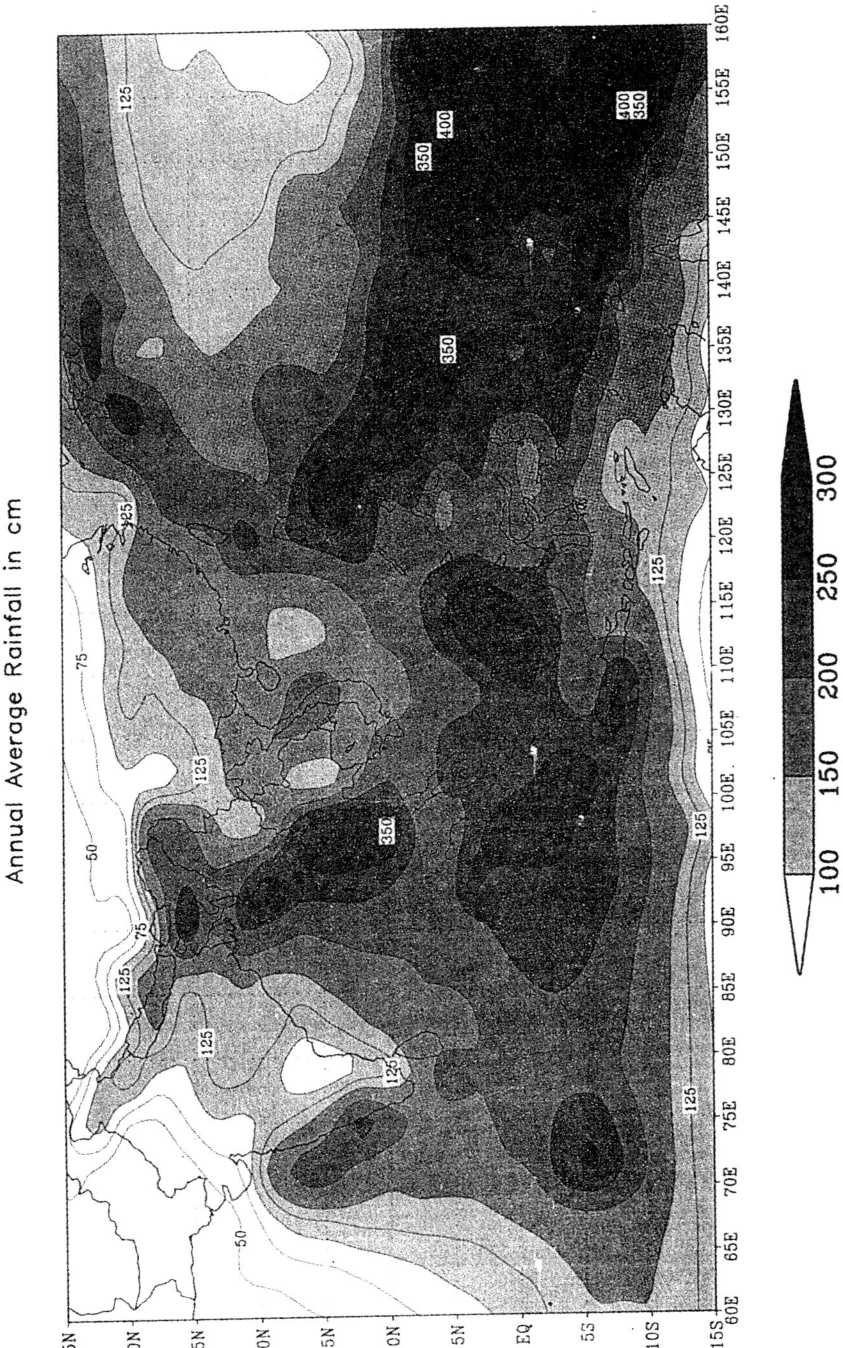

Fig. 6: Average annual rainfall over the Asia-Pacific region (cms).

In this book, we address the problem of understanding the different facets of the links between climate variability and rice production with the ultimate aim of identifying strategies for sustainable production in the face of climate variability. Although we focus on the rice production in India, much of the analysis would apply to other parts of Asia as well.

The top ten states involved in rice production in the country are given in Table 1.

Table 1. Major rice producing states

	1994-95			
State	Area (mha)	% of Total Area	Production (mt)	% of total production
West Bengal	5.8	13.7	12.5	15.4
Uttar Pradesh	5.4	12.8	10.1	12.5
Andhra Pradesh	3.5	8.3	9.2	11.4
Tamil Nadu	2.3	5.5	7.7	9.5
Punjab	2.3	5.4	7.7	9.5
Orissa	4.5	10.6	6.4	7.8
Bihar	4.7	11.2	6.2	7.6
Madhya Pradesh	5.1	12.0	6.0	7.4
Assam	2.5	5.8	3.3	4.0
Maharashtra	1.5	3.6	2.4	2.9

Of these, in West Bengal, Madhya Pradesh and Orissa large fraction of the area under rice is rainfed whereas in Andhra Pradesh, Tamil Nadu and Punjab, the fraction of irrigated land exceeds 0.9. The variation of the rice production, area and yield as well as the rainfall anomaly (departure for the average) for Tamil Nadu and Punjab are shown in Fig. 7 and for West Bengal and Orissa are shown in figure 8. As expected, the variability is smaller for states with a large fraction of irrigated land. It is interesting that until about 1985 the rice production in Tamil Nadu fluctuated from year to year in phase with the rainfall. Subsequently, the production appears to have become relatively insensitive to the vagaries of rainfall. On the other hand, for the states with a large fraction under rainfed conditions the variation in production is large and appear to be linked to the variations in rainfall. It is seen that the major dips in production occur in years of deficit monsoon. Often these are associated with a substantial decrease in the area sown as well as the yield (e.g. 1982 in West Bengal and 1974 in Orissa). In some years there is a decrease in area but not in the yield (e.g. in 1978 in West Bengal when the monsoon rainfall was above normal). This brings out the importance of the distribution of rainfall within the season, particularly that received before transplanting.

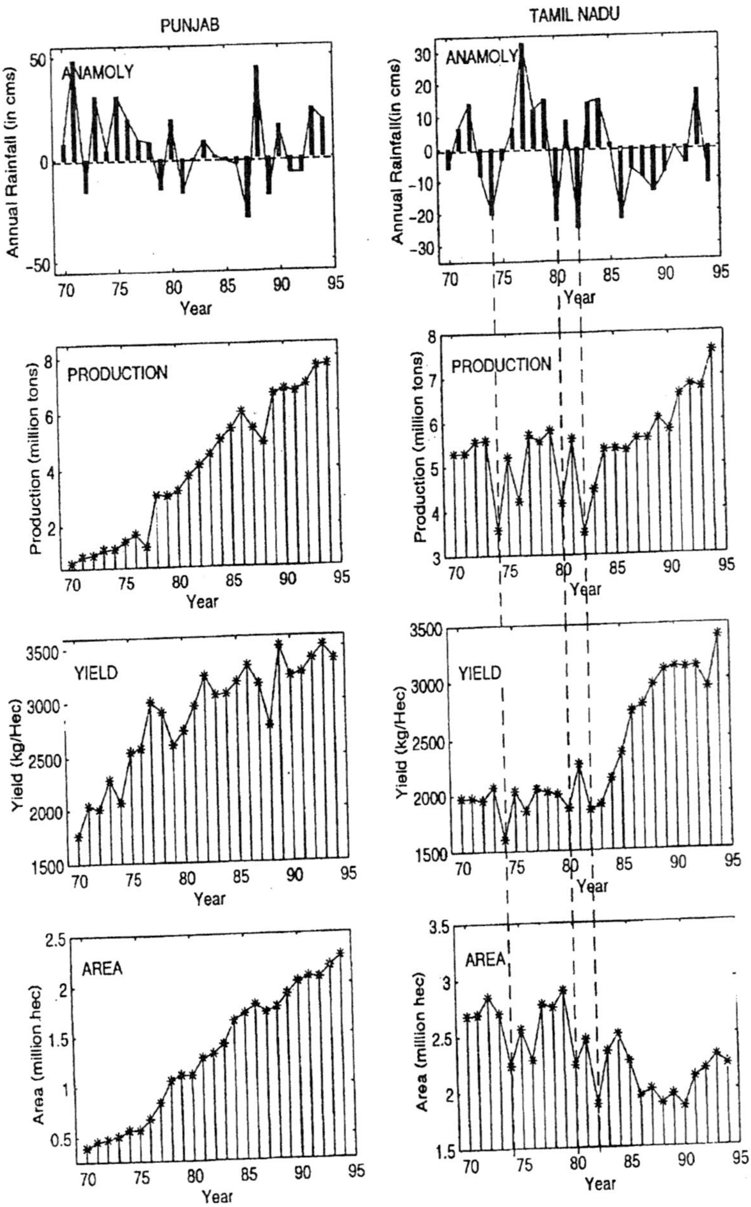

Fig. 7: Variation of the anamoly of annual rainfall (cms), production (million tons), yield (kg/hectare) and area (million hectare) of Punjab (left) and of Tamil Nadu (right) during 1970-1994.

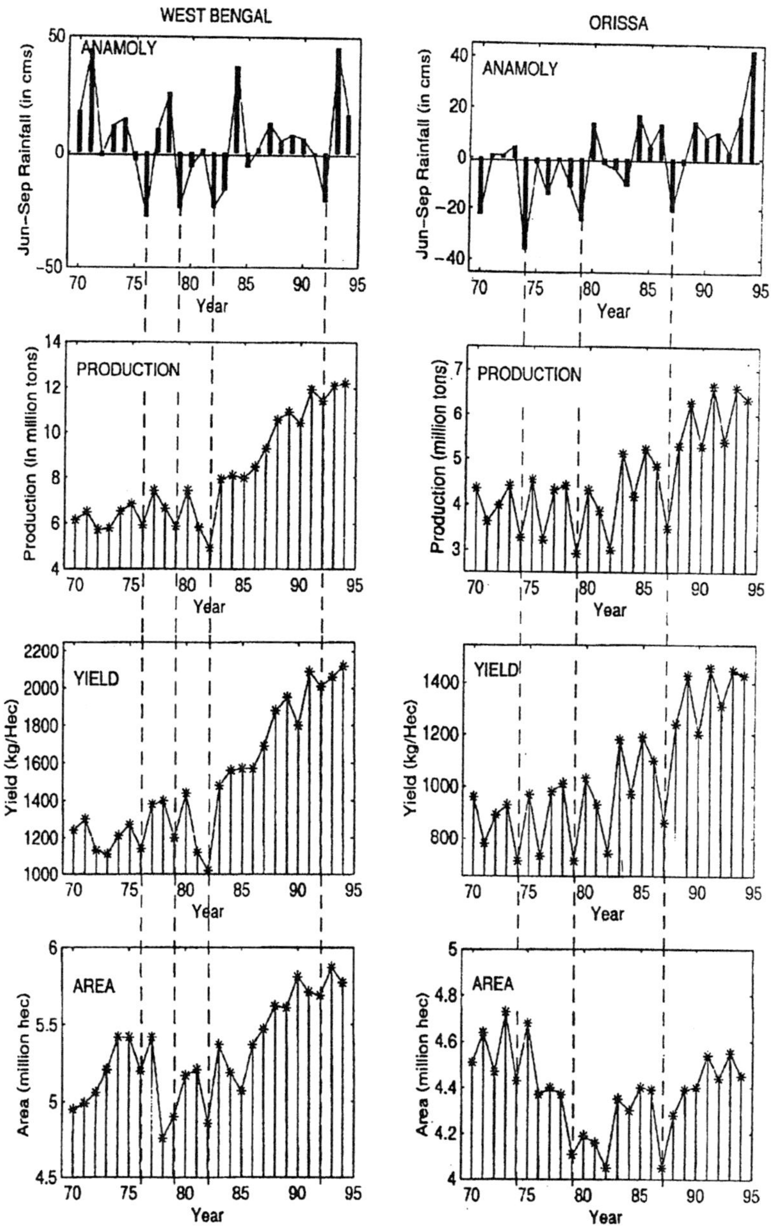

Fig. 8: Variation of the anamoly of June-Sep rainfall (cms), production (million tons), yield (kg/hectare) and area (in million hectare) of West Bengal (left) and of Orissa (right) during 1970-1994.

The major challenge is to determine how to increase the productivity of rainfed areas in the face of the large inter-annual variability of the climate/monsoon. The strategy of introducing new varieties, which was so successful for the irrigated areas during the green revolution, has not worked so well for the rainfed areas. In fact, in Siddiq's assessment (see chapter 7) "it is evident that excessive dependence on the single strategy of varietal solutions to the complex and diverse rainfed ecologies is neither wise nor would be relevant in the coming decades". Given the importance of the climatic resource to these ecosystems, the first task is to address the problem to understanding in depth, the links between climate variability and agricultural productivity, to which little attention was paid during the green revolution operating in controlled environments.

The last two decades have witnessed major advances in our understanding the nature of the variability of the tropical climate, and particularly the monsoon, on time-scales ranging from sub-seasonal to inter-annual and beyond. The spectacular strides in various branches of agricultural sciences over this period have markedly enhanced the possibilities in cropping systems (with new varieties and crops) and management practices (such as application of fertilizers, management of pests/diseases/weeds etc.). There has been considerable development of mechanistic crop models which simulate the entire life history and the productivity of various crops. The productivity of rainfed areas can increase if the major advances in these relevant fields are harnessed for improving our understanding of the nature of the impact of climate on productivity and hence identification of the cropping and management systems which maximize sustained productivity in the face of climate variability.

To attain this goal, a multipronged approach involving analysis of the observed variation of production at different spatial scales (ranging from agricultural stations to district, state, country and even larger), analysis of the variation of the important climatic parameters and development of mechanistic crop models which incorporate realistically the impact of climate and hence can simulate the observed variation in production (such as Figs. 7,8 and smaller spatial scale) reasonably well. Some of the work on the first two aspects is discussed in this part of the book. In the next chapter, the agrometeorology of rice and agroclimatic classification of the country are discussed. Some analysis of the inter-seasonal variation and the available observations of some components of the water balance, particularly evaporation and evapotranspiration over the country are described in the next two chapters. A case study of rice production over a rainfed region of Madhya Pradesh is presented in chapter 5. The prediction of rainfall—which is the most important climatic parameter in the tropics—for the summer monsoon and the post-monsoon (or the northeast monsoon) season over different parts of the country are considered in the last chapter of this section.

References

ABROL, I.P. 1994. Sustaining Indian agriculture—The challenges. In: *Stressed ecosystems and sustainable agriculture*, Eds., Virmani, S.M., J.C. Katyal, E. Eswaran, and I.P. Abrol, Oxford & IBH Publishing Co. Pvt. Ltd., New Delhi. p. 53-64.

ABROL, I.P. 1996. India agriculture scenario. In: *Climate variability and Agriculture*, Eds., Abrol, Y.P., S. Gadgil, and G.B. Pant, Narosa Publishing House, New Delhi, India. p. 19-25.

BLANTRAN DE ROZARI, M. 1995. Rice production and productivity fluctuations in Indonesia. In: *Modelling the impact of climate change on rice production in Asia*, Eds., Mathews R.B., Kropff, M.J. Bachelet D. and Van Laar H.H., CAB International, U.K. p. 41-47.

DEFENG, Z. and SHAOKAI, M. 1995. Rice production in China under current and future climates. In: *Modelling the impact of climate change on rice production in Asia*, Eds., Mathews R.B., Kropff, M.J., Bachelet D. and Van Laar H.H., CAB International, U.K. p. 217-235.

FAGI, A.M. and LAS, I. 1987. Implications of rice-weather studies for national programs. In: *Weather and Rice*—Proc. of International Workshop on the Impact of Weather Parameters on Growth and Yield of Rice. 7-10 April 1986, International Rice Research Institute, Philippines.

RYAN, J.G. 1994. ICRISAT research on sustainable agriculture in the semi-arid tropical stressed environment. In: *Stressed ecosystems and sustainable agriculture*, Eds., Virmani, S.M., J.C. Katyal, E. Eswaran, and I.P. Abrol, Oxford & IBH Publishing Co. Pvt. Ltd. New Delhi. p. 45-52.

SIDDIQ, E.A. 1999. Climate and Rice Production in India. In : *Rice: In a Variable Climate*. (Eds. Y.P. Abrol & Sulochana Gadgil), APC Publications (P) Ltd. New Delhi. p. 105-119.

SURARERKS, V. 1995. Irrigation system management for rice cultivation in Thailand. In: *Modelling the impact of climate change on rice production in Asia*, Eds., Mathews, R.B., Kropff, M.J., Bachelet, D. and Van Laar, H.H., CAB International, U.K. p. 109-127.

VIRMANI, S.M. 1994. Climate resource characterisation in stressed tropical environment: Constraints and opportunities for sustainable agriculture. In: *Stressed ecosystems and sustainable agriculture*, Eds., Virmani, S.M., J.C. Katyal, E. Eswaran, and I.P. Abrol, Oxford & IBH Publishing Co. Pvt. Ltd., New Delhi. p. 149-160.

YOSHINO, M. 1993. Climatic change and agriculture: Problems for the Asian tropics. In: *South-East Asia's Environmental Future—The Search for Sustainability*, Eds. Brookfield, H. and Byron, Y., United Nations University Press. p. 179-189.

Rice—in a Variable Climate, 1999, pp. 25-38
Eds. Y.P. Abrol and Sulochana Gadgil

3. Rice Productivity and Climate

3.1. Perceptions of Rice Climate

Rice is a primary food for most of the Asians and over 85 % of the World rice is produced in Asia. Rice is known as a crop suited to the wet tropics with the highest yields obtained in countries enjoying a subtropical or warm-temperature climate. The crop is grown over a wide range of climatic conditions and soil types including drought prone areas between 45°N and 40°S and over land from sea-level up to 3000 meters above the sea. The water required is high and moderate temperatures and a great deal of sunshine/radiation are considered favourable.

3.1.1. Temperature

The responses of rice to temperature depends on the variety. Crop is adversely affected by high temperature in the lower elevation of the tropics and by lower temperature in the temperate regions. Yoshida (1978) listed critical temperatures at various pheno-phases for the rice crop. Growth and development may be adversely affected if minimum temperature goes below 12°C.

Table 1. Critical temperatures at various stages in the development of the rice crop (Yoshida, 1978)

Growth Stage	Critical Temperature (°C)		
	Low	High	Optimum
Germination	16 - 19	45	18 - 40
Seedling Emergence and Establishment	12 - 15	35	23 - 30
Rooting	16	35	25 - 28
Leaf Elongation	7 - 12	45	31
Tillering	9 - 1	33	25 - 31
Initiation of Panicle Primordia	15	30	-
Anthesis	22	33 - 36	30 - 33
Ripening	12 - 18	30	20 - 29

A few varieties at certain pheno-phase can tolerate temperature as high as 44–45°. Optimum temperature for cultivation varies from 18°C to 33°C. Temperature in the tropics usually remains within this limit.

3.1.2. Radiation

Radiation/duration of bright sunshine hours plays a vital role in rice cultivation. Due to heavy cloudiness during the growing period in the tropics, duration of bright sunshine hours is about 3–4 h/day and yield of rice in this region is low (1.5 – 2.5 t/ha). In the temperate region, yield is about 4–5 t/ha. Duration of bright sunshine hours during monsoon season in the eastern India, Orissa and Andhra Pradesh is of the order of 3–4 hours per day whereas it is around 7–8 hours in the Punjab region. Rice yield in the Punjab region is almost double of the traditional rice growing areas in east and south India. Rice is also now cultivated in summer season in both the regions and duration of bright sunshine hours is about 8–9 hours per day in these areas. Rice yield during summer season becomes almost double of that of monsoon season in eastern and southern India whereas it increases marginally in the Punjab (Biswas, 1996). High yield of 12–14 t/ha has been reported from Pakistan (Rafiq, 1980) where average duration of sunshine observed during growing season is more than 10 hours per day.

3.1.3. Water Requirement

In the humid tropics, where temperature is generally not a limiting factor, the duration of the cropping season depends on rainfall except in irrigated areas. The cropping patterns are often rice-based. In order to describe the various agro-climatic zones in relation to rice-based cropping patterns, water requirements should be determined for rice and for dryland crops. These requirements should then be related to precipitation.

Crop evapo-transpiration (ET) is determined by the climatic environment and the specific crop characteristics. Under conditions of weak to moderate wind speed and medium to high mean relative humidity, ET is determined mainly by radiation intensity. Crop coefficients (Kc) relate ET to PET crop : ET crop = Kc (PET). The value of Kc depends on the crop species and the phenological stage of the crop. It is obvious that less water is required at early growth stages for upland crops than during the mid-season stage. Often, lower amounts of water are needed during the ripening phase. The Kc for most upland crops is between 0.3 and 0.6 during the first month and it is about 1.1 for *bunded* rice throughout the crop cycle. In general, upland crops during the first month require only half the amount of water required at the full canopy stage. ET estimated for legumes and rice during the various phenological stages in both wet and dry seasons is presented in Table 2 (Yoshida, 1978).

Table 2. Crop water requirement for legumes, and rice during four phenological growth stages in the wet and dry seasons (Yoshida, 1978)

Growth Stage	Legumes			Rice		
		Water (mm/mo)			Water (mm/mo)	
	Kc	WS	DS	Kc	WS	DS
Initial	0.50	45	67	1.1	99	148
Development	0.75	67	101	1.1	99	148
Midseason	0.95	85	128	1.15	104	155
Ripening	0.65	58	88	1.0	90	135

Kc = Crop co–efficient ; WS = Wet Season ; DS = Dry Season

The total water requirement also depends on the water losses due to percolation in *bunded* rice. Rice farmers traditionally cultivate their rice soils under wet conditions which lead to formation of low permeable subsurface areas and reduces the rate of percolation. For the alluvial soils, NEDECO (1973) recommends 1 mm water/day percolation. This suggests that water requirement for rice may increase further about 120 mm besides the evapo-transpiration demand.

3.1.4. The Growing Period

The amount of water required to sustain the growth of germinating dryland crops (initial phase) is around half the full rate of evapotranspiration required during the midseason stage. This period is generally not longer than 1 month. Because data on evapo-transpiration are scarce, compared to data on precipitation, the growing period should, if possible, be expressed in terms of rainfall. The relationship between precipitation (P) and the difference between precipitation and evapo-transpiration (P-PET) is shown in Figure 1. Precipitation of more than 120 mm/month may fulfil the evapo-transpiration requirement even in the full growth stage (Oldeman, 1980). The growing period then can be defined as the continuous period (in months) at the onset of the growing period when precipitation is at least 50 mm, plus a period of not more than 1 month at the end of the growing season to allow evapo-transpiration of an amount about 100 mm water stored in the soil profile. This amount of precipitation meets the crop demands.

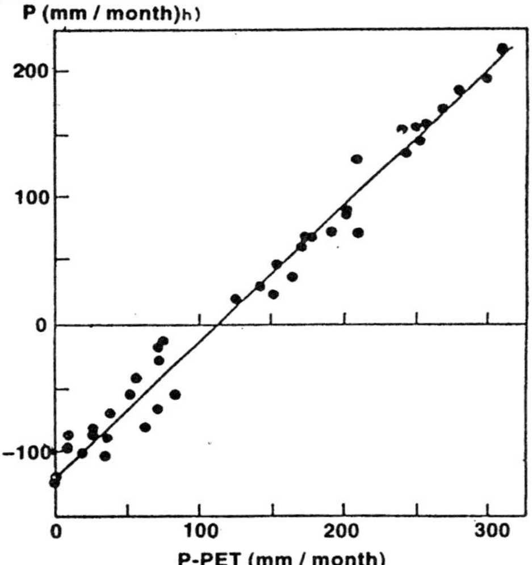

Fig. 1: Relation between precipitation and difference of P-pet

3.2. Climate Regions for Rice

3.2.1. Climate in the Homeland of Domestic Rice

The world map of rice production can be very crudely divided into three types of climate regions. (Figure 2). First, there are the core areas where modest crops are assured year after year with no need for climatic modification by man. This is the "home" area of rice where nature alone controls all of the climate parameters required by the crop. In the homeland (the area of Korat of Thailand, sheltered basin area of northern Thailand, and longitudinal valleys of Mynmar's Shan Upland) of domestic rice, the production season is characterized by considerable rainfall and a great deal of cloud cover. The amount of sunshine received by the crop is surprisingly low, for example the

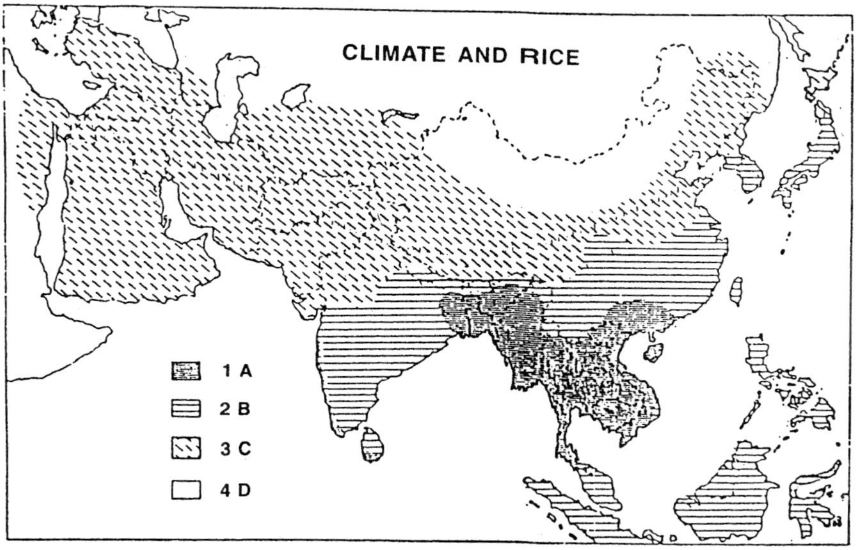

Fig. 2: Climate and rice. 1) "Home" area of rice where the crop may be raised year after year with no climate modification by man. 2) Important rice producing areas where at least once parameter of climate is frequently less than the ideal for successful crops. 3) Areas where rice production is widely scattered and where climate must be modified to produce a crop. 4) Areas with no important production of rice.

mean July sunlight is only 110 hours or about 25 per cent of the potential. The annual solar radiation is only 120 kcal/sq cm, the lowest of any land area in the World at latitude 20° north of south.

Mean monthly temperatures throughout the growing season range from 23.3°C to 27.7°C. In each of the three areas mentioned above, the highest temperatures are recorded during the first half of the season, with slight decrease observed during the ripening and harvest periods. Mean daily maximum temperatures rise to about 33.8°C in the early months at each station while absolute maximum temperatures for the same period reach almost 39°C. During the last half of the growing season mean daily maxima are at the level of 28°C and absolute maxima are only 34°C.

Mean daily minimum temperatures range from 22.7°C to 17.7°C with the lower readings being recorded later in the season, usually well after the flowering period. Absolute minimum temperatures at any of the three locations are recorded at the Mynmar site, at the level of 15°C, and this level is reached only in the final weeks of the season.

In summary, the climate in which rice was perhaps first domesticated, is very warm and humid with a strong monsoon rhythm in the rainfall pattern and with less direct solar radiation than at other areas of the earth. In these areas, rice can be raised without reliance on energy inputs other than for traditional farming practices such as ploughing, planting, and harvesting. The climate is such that a crop can almost always be brought to harvest without the addition of water other than the provided by the rainfall. Temperature ranges are almost never so great as to destroy a crop. The growing season is of ample duration to allow the use of varieties requiring 180 days or more. Rainfall decreases sharply at the end of the season to provide good conditions for the ripening and harvest periods.

3.2.2. Rice Climate in Adverse Situation

At the opposite end of the continuum are the drought prone arid and semi-arid areas such as the Nile Delta, Pakistan's Sindh, northern Hokkaido, and Manchuria (China). In these areas, rice production would be impossible in most years without man's intervention through major engineering works and the expenditure of great amounts of energy. In the Nile and in Sindh, nature is so stingy with her precipitation that rice cultivation (and in fact cultivation of any kind) would be impossible if man did not find ways to provide adequate water. Temperatures are high, relative humidities are low and the potential evapo-transpiration rates are the highest in the World. Water demands for a rice crop in such areas are almost double the needs of the same crop in the core area.

To provide the much needed water, man must intervene and modify the conditions provided by nature. Water is provided by diverting the flow of exotic rivers, themselves out of place in such an environment, and dependent for their flow on moisture provided by nature in sharply contrasting eco-systems hundreds or even thousands of miles away from the irrigated area. Water is moved to the paddy fields through a complex system of dams, reservoirs, canals, siphons and ditches all of which were built and are maintained through the expenditure of vast amounts of energy, chiefly derived from fossil fuels. In other areas, prone to frequent drought and the desert environments, the water is pumped from underground supplies many of which have been accumulated over long periods of time and which are non-renewable in the life of man.

Not only must man provide water in such environments but he must also provide drainage to prevent the accumulation of salts in the surface layers of the soil. Without adequate drainage, the loss of productive capacity can be very rapid, as Pakistan discovered during the 1950s and 1960s and in the Punjab and Haryana in the 1970s. The provision of drainage is still another man-made modification of the environment which required additional energy inputs.

If man is willing to invest sufficient energy, then the hot desert areas and warm drought-prone areas are capable of providing the highest yields per unit of surface obtained from any area in the World. In these areas, high levels of productivity can be maintained year after year with little variation.

This environment is generally not favourable for the pathogens of plant diseases and does not encourage the development of insect pests, Solar radiation is at very high levels and shows little variation from year to year. Clearly, the limiting factor is the availability of the energy needed to provide irrigation and drainage. As long as the food energy derived from the crop exceeds the volume and value of energy inputs, then drought prone and the hot desert areas will remain excellent producers.

Some rice producing areas on polar margins of the rice world have shown remarkable expansion of output in recent decades e.g. Hokkaido and Manchuria. Here the chief environmental or climatic problems are short growing seasons and low temperature, especially early in the season. Again, man through his technological skills and the application of energy derived from fossil fuels has been able to modify the environments so that rice can be produced. Breeding programmes have developed varieties with a shorter growing season, and electric heating cables buried in the soil of the nursery beds have provided increased soil temperatures and protection for the young plants against temperatures too cool for normal plant development. With this use of stored energy and the provision of supplemental irrigation during the long hot summer days, Kokkaido has been converted to an excellent rice-producting area with yields per hectare among the highest in Japan.

3.2.3. Transitional Belt

The third region, and perhaps the most serious problem area, is the great transitional belt between the two extremes. In this broad area map (Figure 2) and including about half of the rice production of Asia and about 45 per cent of the World's rice output. at least one climatic parameter is marginal for rice production, and yet a major portion of the rice crop in this area is produced without micro-climatic modification by man. Most of the drought-prone areas producing rice also belongs to this region. The most serious limiting factor is that much of the crop grown in this region without additional supply of other than that provided by *ponding* of rainfall which occurs naturally in that region. Problems in this zone are legion and include insects, diseases and cost of fuel, but the most critical and least easy to control is the variability over time of climatic parameters, especially rainfall.

3.3. Different Types of Rice

There are many varieties of rice, and the yield of rice also varies widely. Broadly rice may be divided into four types. The nature of agrometeorological information required depends on the kind of rice and the various cultural practices.

3.3.1. Upland

It is grown in fields like corn and other cereals. The field is usually prepared before or at the onset of the rainy season. The rice seeds are usually broadcast or drilled in rows. There are no *bunds* in the field to hold the water, so even distribution of rainfall is very critical for this type of rice culture.

3.3.2. Rainfed

It is raised in *bunded* (diked) fields which collect rainfall. Generally, 10 to 20 day old seedlings are transplanted to the field after ploughing and harrowing it, occasionally, some farmers drill the seeds in furrows when the soil moisture content is at field capacity at the onset of the rainy season. As in the case of upland rice, rainfall distribution is very critical for the success of rainfed paddy.

3.3.3. Irrigated

Rice is grown in *bunded* fields supplies with irrigation water from river or ground sources. In some countries, these rice fields are also provided drainage systems thus ensuring good control of the farm water supply. In Asia, field earmarked to irrigated lowland rice are usually flooded first, then ploughed, fertilized, harrowed, levelled, and transplanted with 10 to 20 days old seedlings. Since the farm water supply is well controlled in this type of rice culture, the distribution of solar radiation during the growing season is more critical than the distribution of rainfall.

3.3.4. Deep Water or Floating

It refers to rice grown in flooded areas, where water depths range from 1 to 6 meters. This type of rice culture is prevalent in some countries of South and Southeast Asia and Eastern and Southern India. Land for this type of rice culture is prepared before or just after the first rain and then the rice are broadcast. Rainfall distribution is also critical to deep water rice especially in the initial stage.

Among the four kinds of rice culture, irrigated lowland rice gives the highest yield, followed by rainfed paddy, upland rice and deep-water rice. Two to three crops of irrigated rice can be produced per year. Farmers continue to plant deep-water rice although yields are low since no other food crop can be grown in flooded areas.

3.4. Agro-climatic Classification of the Indian Region for Rice

In the north-western parts of the country region, rainfall is so low and so variable from year to year that rice is raised exclusively with irrigation. This corresponds to the adverse situation discussed in 3.2.2. In the Ganges Delta, Assam, Bengal, eastern parts of Bihar and along the western coastal lowlands, rainfall is sufficiently heavy and reliable to provide only modest variation in output. This area is 'home' of rice as discussed in 3.2.1. However, in the central India from Uttar Pradesh and western Bihar in the north to Tamil Nadu and Andhra Pradesh in the south, the mean rainfall totals are marginal for rice and the variability from year to year both in the timing of the onset of the rains and in the total volume of precipitation is very great. A large portion of the area under rice is either unirrigated or watered from unreliable sources such as tanks or shallow wells. This is the transitional belt in which the major portion of areas of frequent droughts falls.

As adequate high temperature is prevailing in the tropics throughout the year, most of the classification attempted are based on precipitation and soils. Mohr (1933) proposed the first classification for the humid tropics, which was based on the number of wet and dry months. Vander Eelaart (1973) prepared an agroclimatic classification for Thailand based on the number of consecutive humid months. Panabokke (1979) developed a map of agro-climatic zones of south east Asia and divided the area into six zones. This was based on intensity of rainfall and its variability during the wet season. Nieuwolt (1981) proposed an agro-climatic rainfall index (ARI) based on rainfall data as percentage of potential cvapo-transpiration:

$$ARI = 100\ XP/PET$$

where P and PET are monthly rainfall and potential evapo-transpiration, respectively. He used monthly rainfall at 80 percent probability level instead of monthly mean rainfall and divided Malaysia into various agro-climatic zones.

The International Rice Research Institute (1974) introduced an agro-climatic classification to give a representative spectrum of rice growing environment of Southeast Asia. These zones were based on wet, dry and intermediate months. A month is defined as wet month when mean rainfall exceeds 200 mm, dry when rainfall is less than 100 mm and intermediate if monthly rainfall lies between 200 mm and 100 mm. This classification has brought out five main zones based on wet months viz. Zone A : more than 9 consecutive wet months ; Zone B : 7–9 consecutive wet months ; Zone C : 5–6 consecutive wet months ; Zone D : 3–4 consecutive wet months and Zone E : less than three consecutive wet months. These zones are further sub-divided according to the duration of dry and intermediate months.

3.5. Classification Involving Risk Factor

The studies mentioned so far do not deal with the risk factor involved in any planning. It goes without saying that the planner, be a farmer or a scientist, must know the risk he is going to take in his endeavour. The study using average or normal rainfall cannot include this factor. Even in high rainfall areas, there is considerable year to year deviation of rainfall from the normal. In such areas, the normal rainfall (monthly, seasonal or annual) may be adequate to raise the crops. But experience shows that on a number of occasions crops cannot be raised. Apparently, in such years the rainfall is less than the normal value and cannot meet the water demand of the crops. So, for any crop planning purpose, one should know from past records what is the chance of meeting the water requirement of the crop. Accordingly, the planning has to be done on a probabilistic basis, which eventually takes into account the chance of success or failure.

Table 3. Rainfall at different probability level

WEEK	22	23	24	25	26	27	28	29	30	31	32	33	34	35	36	37	38	39	40
									CUTTACK										
AVG	23	37	53	60	80	73	82	67	88	96	82	77	81	64	65	60	46	48	41
30	20	44	65	69	91	90	91	89	116	106	91	86	90	83	100	79	56	44	42
50	16	25	37	45	60	51	61	49	63	71	60	52	60	51	54	43	37	27	34
70	6	11	23	28	37	32	38	31	41	45	39	32	38	35	35	28	23	16	11
									JHANSI										
AVG	7	7	13	24	43	47	61	91	73	71	93	68	54	53	49	38	38	22	8
30	5	6	15	29	59	55	76	109	84	81	108	81	60	64	64	52	48	22	6
50	0	0	6	16	26	32	44	64	56	48	64	47	37	37	35	22	15	3	0
70	0	0	0	7	8	15	26	41	35	24	31	22	23	16	11	6	0	0	0
									SILCHAR										
AVG	108	164	153	134	126	127	123	126	115	113	117	120	105	110	94	95	89	73	72
30	125	161	176	164	150	148	143	157	151	143	144	137	132	137	120	111	103	86	81
50	74	116	132	121	111	110	108	116	94	105	106	103	97	102	75	83	61	51	49
70	48	73	89	80	85	75	68	59	68	68	62	69	56	59	51	56	39	32	30
90	14	34	42	41	39	38	32	27	33	32	30	34	28	27	24	22	19	17	8

The weekly average rainfall at 30 % , 50 % and 70 % probability levels of three stations, where rice is grown during monsoon season is given in Table 3. It is seen at Cuttack that weekly average is adequate to raise rice crop every year. Rainfall at 70 % level also indicates that rainfed rice can be cultivated without irrigation but at higher probability such as (85–90 %) only a dry land can be raised. In case of Jhansi, rainfed rice cannot be grown at 70 % level, but a dry land crop of 12–14 week duration can be harvested from this station. Silchar is situated in high rainfall area. Even 90 % is more than adequate to raise rice crop from this station. Two rice crops can be grown with proper management of rainfall.

Hargreaves (1971) proposed a classification using monthly probabilistic rainfall at 75 % probability level instead of the normal rainfall. He used the incomplete gamma distribution to compute such values. He classified the agroclimate of NE Brazil, a semi-arid region on the basis of monthly moisture. availability index. This is an important step because planning has been done considering chances of failure or success of an endeavour. It is obvious that the action taken by the planner will depend heavily on his choice of the risk level.

Biswas (1982), and Sarkar and Biswas (1986) improved Hargreaves's classification by introducing modifications. This has been further modified to identify rice growing areas. The classification is as follows: (I) Weekly MAI rather than monthly has been used which helps in getting the appropriate effect of distribution of adequate amount of rainfall (ii) different risk levels instead of one so that the planners can choose their own risk level and (iii) MAI > 0.5 and MAI > 1.0 depending upon the crop-growth phase. In this study, moisture availability index was defined as follows:

$$\text{MAI} = \frac{\text{Weekly Assured Rainfall at 50\% Probability Level}}{\text{Potential evapo-transpiration}}$$

MAI = Moisture Availability Index

Criteria for the present classification are as follows:

Classification	No. of Week	When MAI at 50% Level
	0.5	1.0
A	≥ 13	≥ 7
B	≥ 18	≥ 12
C	≥ 22	≥ 15
D	≥ 28	≥ 21

Assured rainfall required to compute MAI has been obtained by fitting incomplete Gamma distribution to historic weekly rainfall data.

On the basis of weekly MAI at 50 % level, the climate was divided into 4 major groups which were further sub-divided on the basis of duration of mid-season water stress period. Considering the values of MAI at all the levels, a suitable level for crop-growing can be selected for each zone depending upon the duration of adequacy of moisture.

In many areas, especially in the tropics, rainfall is bi-modal, which causes water stress after 4/5 weeks of sowing. In this classification, mid-season water stress period, when MAI is less than 0.5 has been considered. Water stress period of various duration is also indicated by using numerical suffixes in the main classification.

Cessation of rainy season does not mean the end of the crop season. Crops can thrive on stored moisture. This has been taken into account considering the accumulated assured rainfall and potential evapo-transpiration. Studies conducted by different scientists (Holmes and Robertson, 1963, Riplay, 1966, Jenson, 1968; Biswas and Khambete, 1987) show that water consumption by dryland crops like sorghum, millet etc. during the whole life span is about 60–70 per cent of potential evapo-transpiration. It has been stated earlier that crop coefficient for rice at its different phenophases remains around one. Hence it is assumed that crop requires water equivalent to potential evapo-transpiration during its life span. It has, therefore, been considered in this study that the difference between the accumulated assured rainfall and total PE of the same period plus percolation will go into soil as stored moisture and the plant can use it even after the end of rainy season. Better results can be achieved by super-imposing this classification on soil map of the region to bring out crop potential in terms of climate and soil.

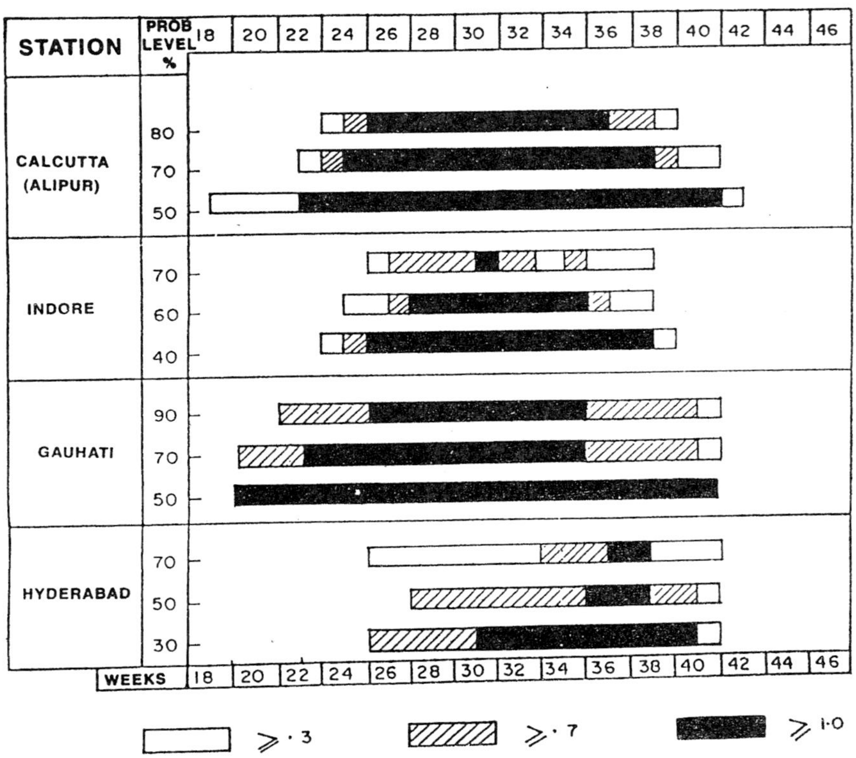

Fig. 3: Moisture availability index for crop growing period

Crop potential of 4 stations situated at different parts of India, where rice is one of the major crops, is depicted in Figure 3. It appears that rice can be raised throughout the year at Guwahati. Even two

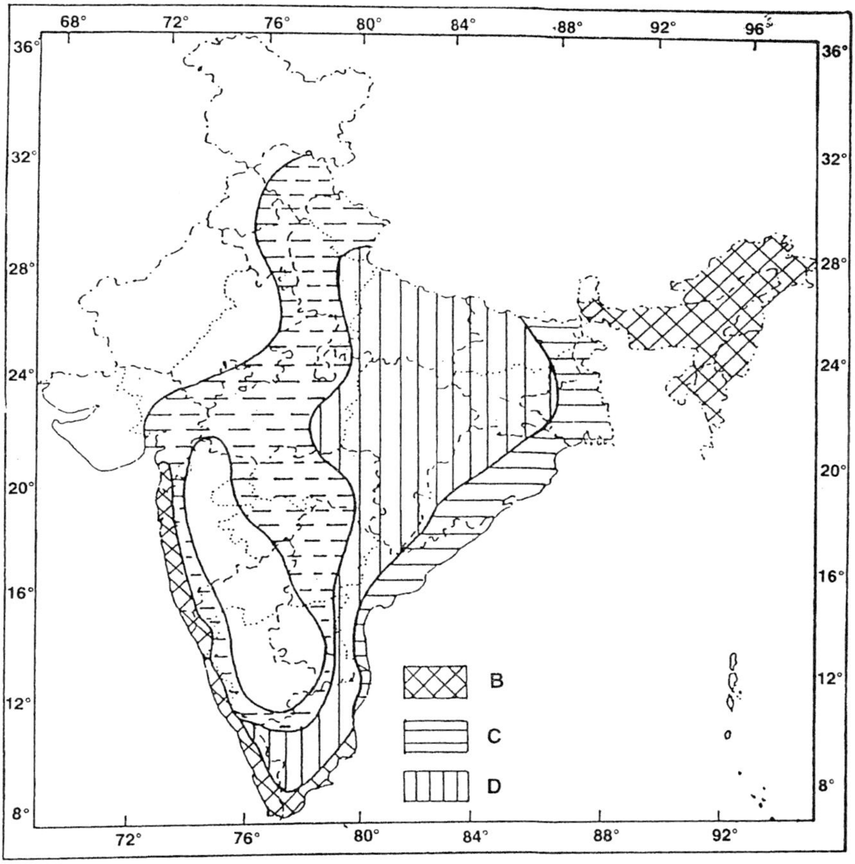

Fig. 4: Agroclimatic zones for rice

crops can be raised with proper management. At Indore, it can be grown only 6–7 years out of 10 years. Under rainfed condition, rice can be raised from Hyderabad region at 40 % probability level. Rice can be harvested from Calcutta region in nine out of ten years. Different rice growing areas in India are given in Figure 4. From area C rice can be grown every year. In the Assam region, even two crops can be raised by efficient management of water. From area B, rice can be grown 80–85 % of the year and from area A it can be harvested 60–70 per cent probability level under rainfed condition.

3.6. References

BISWAS, B.C. 1982. Agroclimatic classification on the basis of moisture availability index and its application to dry farming tract of Gujarat, *Mausam.* **33**: 465-476.

BISWAS, B.C. 1996. Link between climatic variability and agricultural production. In: *Climate variability and agriculture.* (Eds. Y.P. Abrol, S. Gadgil and G.B. Pant) Narosa Publishing House, New Delhi. p. 188-203.

BISWAS, B.C. and KHAMBETE, N.N. 1987. Water consumption by dryland crops as related to pan evaporation. *Mausam*. **39**: 91-96

HARGREAVES, G.H. 1971. Water requirements manual for irrigated crops and rainfed agriculture. Pub. No. 75-D-159. EMBRAPA and Utah State University.

HOLMES, R.M. and ROBERTSON, G.W. 1963. Application of the relationship between actual and potential evapotranspiration in dryland agricultural transaction. *American Soc. Agri. Engi.* **6**: 65-67.

INTERNATIONAL RICE RESEARCH INSTITUTE, 1974. An agroclimatic classification for evaluating cropping system potentials in southeast Asia rice growing regions, IRRI Report, Los Banos, Philippines. p. 1-5.

JENSON, M.E. 1968. Water consumption by agricultural plants. *Water deficit and plant growth*, Vol. 2, Academic Press. New York, pp. 1-22.

MOHR, E.C. 1933. De boden der tropen inhet algemeen en die van Ned. India in het byzonder (in Dutch). Mededelingen Kon. Ver. Kol. Inst. No. 31, Vol. I, Amsterdam.

NIEUWOLT, S. 1981. The agricultural rainfall index (a mimeograph), MARDI, Agrometeorological Unit, Malaysia, p. 13.

OLDEMAN, L.R. 1980. The agroclimatic classification of rice growing environments in Indonesia. In: Proc. Agrometeorology of the rice crop held at IRRI.

PANABOKKE, C.R. 1979. Agroecological zones of south and southeast Asia. FAO regional office for Asia and in the Far-East, Bangkok, Thailand.

RAFIQ, M. 1980. Agrometeorology of three rice regions of the Indus Plain. Proc. Agrometeorology of rice crop held at IRRI.

RIPLAY, P.O. 1966. The use of water by crops. In: Proc. ICID special session: *New and Promising developments in the field of irrigation drainage and flood control*—January, 1966, New Delhi. International Commission on Irrigation and Drainage.

SARKAR, R.P. and Biswas, B.C. 1988. New approach to agroclimatic classification to find out crop potential. *Mausam*, **39**: 343-358.

TROLL, C. 1965. Seasonal climates of the earth, In: *World Maps of Climatology* (Ed. E. Rodenwalt and H. Juretz) Springer-Verlag, Berlin, p. 28.

VANDER EELAART, 1973. *Climate and crops in Thailand.* Soil Survey Division, Min. Agric. and Coop., Dept. of Land Dev. and FAO Report SSR-8, Bangkok.

YOSHIDA, S. 1978. Tropical climate and its influence on rice IRRI. Res. Paper Ser. 20 Los Banos, Philippines.

Rice—in a Variable Climate, 1999, pp. 39-50
Eds. Y.P. Abrol and Sulochana Gadgil

4. Intra-seasonal Variations and Crop Planning in Drought Prone Areas

Planning of agricultural operations in any region particularly in drought prone areas requires an understanding of the rainfall characteristics at each rainfall station in greater detail. Frequency distribution of daily, weekly and monthly rainfall of specified amounts, and evaluation of their probabilities of occurrences, frequency of dry spells of various lengths and their probabilities, chances of a dry or wet week(s) based on rainfall situations in preceding week(s) etc. are essential prerequisites for judicious use in water management and crop planning. In no other region and crop this is as important as in the drought prone areas and for rice crop which needs largest amount of water among cereal crops.

Many criteria have been suggested in the literature for the study of rainfall reliability in relation to meteorological and agricultural drought and their effect on crop growth and yield. Probabilistic approach have been in vogue to study reliability of rainfall and determining drought proneness. Barger and Thorn (1949) advocated use of threshold amount of rainfall needed for crop growth to maturity. Chowdhury and Abhyankar (1979) introduced threshold values of cumulative rainfall necessary to give yield above normal. Use of evapotranspiration (ET) and soil water at different stages have been suggested by Van Bavel (1953) as rainfall reliability factors in relation to crop growth. With high ET rates and low soil moisture storage, Van Bavel's approach ideally suited in areas prone to droughts.

4.1. Evapotranspiration Measurements in Rice Crop

For efficient water and crop (rice) management, accurate information of the evapotranspiration needs of the crop is a must.

Agricultural Meteorology Division has in the course of time since 1973, established an efficient network of four Evapotranspiration Measurement Observations all over India covering different agroclimatic zones. Observations are recorded on almost all the crops grown in the area. There are 34 gravimetric lysimetric observatories and 7 volumetric lysimetric observatories. Apart from measurements on evapotranspiration, soil moisture observations in the cropped and crop free (bare) field are also observed once in a week. These lysimetric observatory locations are so chosen that simultaneously other meteorological parameters are also available from nearby crop weather observatory (C.W.O.) maintained and manned by the co-operating agencies viz. Agricultural Universities, Agricultural Experimental farms of ICAR, State Agricultural Departments and other such agencies engaged in agricultural research programmes.

The seven volumetric lysimetric stations are Bikramgunj in Bihar, Canning in West Bengal, Pattambi in Kerala, Raipur in Madhya Pradesh, Samakhunta in Orissa, Annamalainagar in Tamil Nadu and Nellore in Andhra Pradesh. Nellore and Annamalainagar receive North East Monsoon rains in winter and ramaining five stations receive southwest monsoon rains in wet summer season and accordingly, the rice crop growing seasons also differ.

ET observations are recorded daily alongwith crop observations. Relevant observations in crop are recorded so as to demarcate different phenophases, attack of any pest and disease or any stresses etc. Harvesting is done separately for field and the lysimeter crop.

Weekly ET data for the rice crop grown in the year 1995 at Raipur is presented alongwith simultaneous meteorological conditions experienced by the crop. (Table 2) Crop remained in the field for 123 days with average daily ET of 4.48 mm with maximum water utilised (50 %) at its vegetative phase and only 33 % at maturity phase of the total ET loss of 551 mm. Average daily ET loss of 5.68 mm was observed during flowering. Water use efficiency was found to be 9.27 kg ha^{-1} mm^{-1}. Temperatures at tillering about 27°C. at flowering 26° to 27°C were very favourable for the crop performance. Average crop coefficient at flowering stage was maximum 1.14 indicating water loss was more than atmospheric demand.

Table 1. List of E.T. stations

No.	Name of the station	State	Lat.	Long.	Starting date
1.	Akola	Mah.	20°42'	77°02'	July '73
2.	Anand	Guj.	22°35'	72°55'	Jan. '78
3.	Anakapalle	A.P.	17°38'	83°01'	April '74
4.	Annamalainagar (V)	T.N.	11°24'	79°44'	Aug. '76
5.	Bangalore	Kar.	13°00'	77°37'	Oct. '76
6.	Banswara	Raj.	23°33'	74°27'	Dec. '78
7.	Bari-Brahmana	J & K	32°39'	74°54'	Dec. '78
8.	Bellary	Kar.	15°09'	76°51'	July '74
9.	Bhubaneswar	Orissa	20°15'	85°52'	June '78
10.	Bikramganj (V)	Bihar	25°10'	84°05'	Oct. '77
11.	Canning (V)	W.B.	22°15'	88°40'	July '75
12.	Coimbatore	T.N.	11°00'	77°00'	Aug. '73
13.	Dharwar	Kar.	15°26'	75°07'	June '78
14.	Hisar	Haryana	29°10'	75°46'	Dec. '76
15.	Jabalpur	M.P.	23°09'	79°58'	June '76
16.	Jhansi	U.P.	25°26'	78°35'	Dec. '78
17.	Jodhpur	Raj.	26°18'	73°01'	May '75
18.	Jorhat	Assam	26°47'	94°12'	June '76
19.	Kovilpatti	T.N.	09°12	77°53'	Aug. '76
20.	Lucknow	U.P.	26°52'	80°56'	Feb. '79
21.	Ludhiana	Punjab	30°56'	75°52'	June '77
22.	Nellore (V)	A.P.	14°27'	79°59'	Aug. '75
23.	New Delhi	Delhi	28°40'	77°10'	May '75
24.	Pantnagar	H.P.	29°.00'	79°30'	July '77
25.	Parbhani	Mah.	19°16'	76°47'	Dec. '79
26.	Pattambi (V)	Kerala	10°48'	76°12'	June '78
27.	Rahuri	Mah.	19°24'	74°39'	July '77
28.	Raipur (V)	M.P.	21°16'	81°36'	Aug. '77
29.	Rajahmundry	A.P.	17°00'	81°46'	Mar. '78
30.	Rajendranagar	A.P.	17°14'	78°33'	April '78
31.	Rajkot	Guj.	22°17'	70°48'	April '78
32.	Ranchi	Bihar	23°17'	85°19'	Aug. '76
33.	Samakhunta (V)	Orissa	21°56'	86°46'	Nov. '80
34.	Solapur	Mah.	17°04'	75°54'	Sept. '76
35.	Varanasi	U.P.	25°18'	83°03'	July '78
36.	Bhopal	M.P.	23°16'	77°25'	Dec. '89
37.	Dantiwada	Guj.	24°12'	72°30'	Jan. '90
38.	Durgapura	Raj.	26°51'	76°47'	Nov. '89
39.	Tirupati	A.P.	13°05'	79°05'	May '90
40.	Patanchuru (ICRISAT)	A.P.	17°32'	78°16'	April '78
41.	Pune	Mah.	18°32'	73°51'	June '73

Information on year wise evapotranspiration measurements at Pattambi and Raipur, wherein variety cultivated, crop duration,

Table 2. Weekly BT data and crop phenological stage

CROP : PADDY (MAHAMAYA) **TRANSPLANTATION : 26.07.1995** **HARVESTING : 26.11.1995** **STATION : RAIPUR**

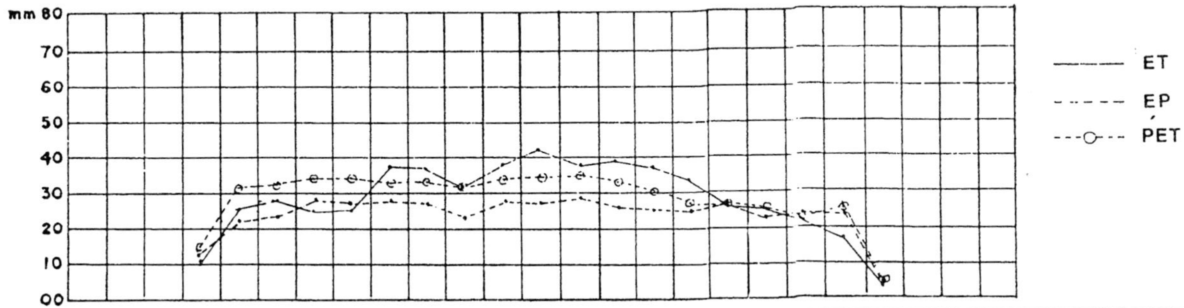

		30	31	32	33	34	35	36	37	38	39	40	41	42	43	44	45	46	47	48	Season's Total
Weekly Total (mm)	ET	10.2	25.7	29.4	25.4	26.3	38.9	38.0	32.4	39.8	41.7	38.3	39.9	37.2	33.0	26.4	26.3	22.1	18.3	01.8	551.7
	EP	11.5	22.8	24.2	28.9	27.8	28.9	27.4	23.4	28.9	28.5	29.6	26.4	26.0	25.0	27.0	23.9	24.4	24.7	02.9	462.2
	PET	15.0	31.5	32.2	35.0	35.0	32.9	33.6	32.2	34.3	35.0	35.7	32.9	30.1	27.3	27.3	26.6	23.8	25.2	03.2	548.8
	Rainfall	02.0	61.8	78.8	151.0	16.4	111.2	04.8	03.0	21.2	13.0		02.6	25.8			08.2				499.8
Rainy Days			2	4	4	1	6	–	1	2	1		–	3			1				25.0
Crop Coefficient		0.68	0.81	0.91	0.73	0.75	1.18	1.13	1.01	1.16	1.19	1.07	1.21	1.23	1.21	0.97	0.99	0.93	0.73	0.56	
Mean temp. °C	Max.	31.5	30.1	30.5	31.0	31.1	31.2	30.3	32.1	31.7	31.8	32.7	30.7	30.6	29.5	29.6	28.3	29.8	29.2	28.4	
	Min.	25.1	24.7	24.1	24.2	24.6	24.8	24.2	24.8	24.3	23.7	22.6	22.4	22.5	17.6	16.1	14.8	12.2	16.1	13.4	
Meteorological Week		30	31	32	33	34	35	36	37	38	39	40	41	42	43	44	45	46	47	48	
Growth Stage		← (i) →										← (ii) →		← (iii) →							
Month/Period			Aug/VIII					Sep/IX				Oct/X					Nov/XI				

Stagewise Evapotranspiration (mm)		:		Irrigation Dates
i) Vegetative Stage	278.0	:	(50 % of Total ET)	-------------------
ii) Flowering Stage	90.9	:	(17 -------- " ------)	Not Applicable
iii) Grain Formation Stage	182.2	:	(-------- " ------)	-------------------
iv) Maturity Stage		:	(33 -------- " ------)	-------------------

Yield (Q/hec)	
Tank	51.10
Field	48.60
Water Use Efficiency	9.27 kg/ha/mn

Table 3. Evaporation measurements variety caltivated, crop duration, phasewise water consumption, crop coefficients

Year	Variety	Duration (days)	ETd^{-1} (mmd^{-1})	Phasewise water consumption (mmd^{-1})			Phasewise crop coefficients			WATER USE EFFICIENCY (Kg. ha^{-1} mm^{-1})
				V	F	M	V	F	M	
STATION : PATTAMBI										
1978	JAYA	106	5.61	5.02	6.09	4.68	1.29	1.42	1.39	9.77
1979	JAYA	124	4.80	5.35	5.40	4.97	1.28	1.02	1.11	12.84
1980	JAYA	91	4.64	4.40	4.93	5.71	1.11	1.00	1.03	9.11
1981	TRIVENI	84	4.45	3.81	4.92	6.73	0.89	1.05	1.43	13.89
1982	DO	87	4.45	4.11	5.61	2.77	1.05	1.02	0.99	16.09
1983	DO	90	4.44	—	—	—	—	—	—	12.28
1984	DO	95	6.65	5.79	7.95	7.71	1.35	1.78	1.55	4.17
1985	DO	89	4.77	3.57	5.86	5.42	0.99	1.26	1.13	5.72
1992	TRIVANI RED	90	5.11	4.98	6.33	4.75	1.29	1.44	1.31	8.08
STATION : RAIPUR										
1977	BD-817	87	4.44	—	—	—	—	—	—	—
1978	BD-817	83	3.36	3.86	3.20	258	—	—	—	10.20
1979	DO	100	5.61	—	—	—	—	—	—	3.40
1980	JR-155-355	106	6.07	6.60	5.71	4.50	1.34	1.37	1.09	7.02
1981	R-2384	116	6.33	7.53	7.23	4.37	1.31	1.47	1.44	2.79
1982	R-2384	108	6.55	6.03	8.98	6.61	1.27	1.74	1.46	4.05
1983	R-2384	105	4.56	4.45	4.67	4.73	0.99	1.04	1.09	4.35
1984	R-2384	114	6.77	6.89	6.97	6.54	1.38	1.31	1.55	4.23
1985	R-243/3223	113	4.16	4.31	4.85	3.23	0.99	0.97	0.78	3.34
1990	R-289/WR-3712	109	6.44	6.30	8.46	6.30	—	—	—	6.68
1991	IR-36	122	3.36	4.44	5.62	3.57	0.97	1.12	0.94	12.50
1992	MADHURI	111	5.77	6.21	6.44	4.73	1.34	1.37	1.27	7.59
1993	IR-64	108	5.39	6.01	7.38	4.79	—	—	—	11.71
1994	KRANTI	125	4.66	4.74	4.89	4.44	1.07	1.09	1.22	7.89
1995	MAHAMAYA	123	4.48	4.56	5.68	3.96	0.95	1.14	1.03	9.27

average daily ET loss for the entire crop duration, phasewise water consumption (ET) for the three broad phases (vegetative, flowering and maturity), Kc values for the three broad phases and water use efficiency are presented. (Table 3).

4.2. Wet and Dry Spells

It is well known that consecutive spells of low rainfall invariably result in drought conditions and poor yields. Successive high-rainfall days also result in flooding, water logging and lodging. Use of probabilities of dry spell runs gives the indication of the extent of the moisture stress and agricultural drought experienced by crops during the rainy or south-west monsoon season (Sastry, 1976; Das and Datar, 1996).

Results of analysis of dry spells during the southwest monsoon season from July-October at New Delhi (28° 35'N, 77° 12'E) are given in Tables 4 and 5, which show the frequency and probability of occurrence of a specified number of consecutive dry days, respectively.

Table 4. Frequency of continuous dry days (rainfall less than 6 mm/24 hr) during the weekly periods of July-October (based on the data for the period 1940-1972), station, New Delhi

Station: New Delhi	Frequencies in 33 years																
No. of continuous dry days in week	Week Nos (first week ends on 7th July. 17th week ends on 27th Oct)																
	1	2	3	4	5	6	7	8	9	10	11	12	13	14	15	16	17
Four	29	20	17	20	22	17	19	23	26	23	26	28	31	32	32	32	33
Seven	16	10	07	09	07	11	06	08	14	16	13	21	21	25	29	29	32

Table 5. Probability of occurrence of specified number of continuous dry days (rainfall in 24 hr less than 6 mm) at New Delhi.

Month	No. of continuous dry days (Less than 6 mm per day)	Probable frequency
July	07	Once in 3 years
	20	Once in 6 years
August	07	Once in 5 years
	20	Once in 15 years

For the purpose of this analysis, a "dry" day has been defined as a day with rainfall less than 6 mm in 24 hours, which is based on the fact that the average potential evapotranspiration rate is about 5 mm/24 hr at the station for the season. It has been shown that a crop which cannot withstand stress for at least 4 consecutive days is likely to be adversely affected in this environment. Frequency distribution of 7-day dry periods shows that between the weeks 3 to 8, which usually corresponds to the vegetative growth phase of crops at New Delhi, week number 6 is relatively dry. It can be inferred from the data that prolonged dry spells of 20-day duration are more frequent in July than in August.

By carrying out similar analysis with respect to the different growth phases, the periods that are likely to be subjected to stress under different rainfall situations at a particular location can be identified using past rainfall data (Table 6). These results show that the severity of drought depends on the distribution of rainfall and the phases of crop growth affected by dry periods, and not on the total amount of rainfall alone (Anonymous, 1970).

Table 6. Distribution of dry weekly periods, starting July 1st in the crop phases of maize at New Delhi : (i) sowing and germination (ii) growth period (iii) tasselling period (iv) grain filling and mature stage.

Year	Week No.																	x Dry week	
	1	2	3	4	5	6	7	8	9	10	11	12	13	14	15	16	17	Rainfall (Jul-Oct. mm)	Drought Class
1951		x	x	x		x		x	x	x		x	x	x	x	x	x	214	Disastrous
1940	x	x					x		x	x	x	x		x	x	x	x	527	Severe
1952		x								x	x	x	x	x	x	x	x	360	Moderate
1942		x				x						x		x	x	x	x	522	Light
1961	x			x							x			x		x	x	925	Unaffected
		i			ii					iii		iv							

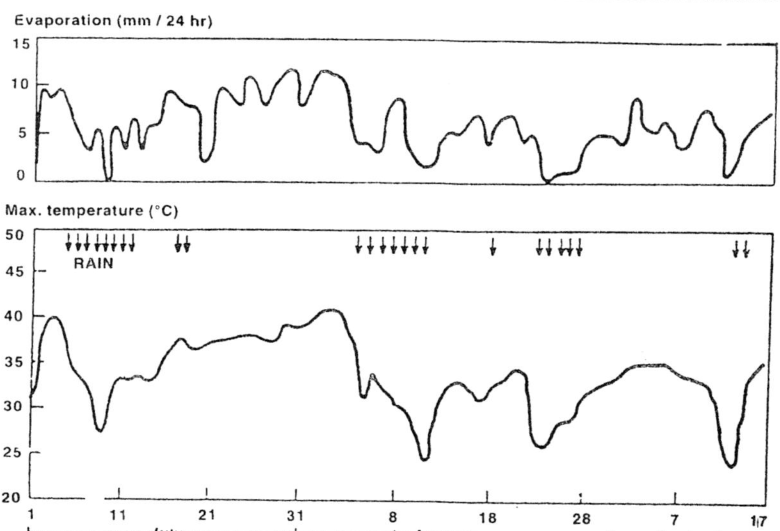

Fig. 1: Rainy days, class A covered pan evaporation and maximum temperature during *kharif* 1972 at New Delhi (Farm area, IARI)

The choice of continuous dry days to represent drought has this advantage: the non-occurrence of rainfall during the rainy season on a number of consecutive days in the tropics and sub-tropics especially in drought prone areas is associated with prevalence of high temperatures and high evaporation which indicate dessication of atmosphere and plant stress conditions.

4.3. Dry Weeks and Bright Sunshine

Cloudy and rainy days are detrimental to the rice crop, particularly from flowering to maturity. Information on the relative probability of getting dry week with sufficient bright sunshine aids in determining the best flowering-to-maturity period of the crop at the station.

Similarly, by extending the analysis over a network of rain gauge stations, choosing a criterion of rainfall for defining consecutive wet days, the assured rainfall periods for supplementary irrigation can be ascertained, at the threshold value chosen for any crop; under consideration. Different criteria should be adopted in different stations of drought prone areas where soil type and hence water holding capacity vary even at short distances. The results may then be integrated over the entire drought-prone area.

Thus, depending on the nature of the problem, a judicious choice of threshold or limiting values of the amount of rainfall and duration of dry or wet spells for analysing rainfall data could provide information useful in crop planning. This is not to suggest that other climatic parameters should not be considered. Where extensive and reliable data on various climatic elements are available, each parameter may be analyzed, using more or less sophisticated methods, to derive quantitative results.

4.4. Crop Planning and Varietal Characteristics in Drought Prone Areas

One of the main objectives of crop planning is to identify the areas where a particular crop could be grown successfully under given climatic conditions. Nearly every crop variety responds in its own way to the atmospheric environment at a location and the same climatic element can have quite different effects at even on the same variety, depending on its stage of growth. Varieties also may differ in their ability to withstand stress in different stages of growth. For example, rice varieties like *Cauvery* wilt at moderate drought but rejuvenate better than most other varieties after re-irrigation (Balakrishnarao *et al.*, 1971).

To derive maximum benefits from each of the several varieties of a single crop species by its introduction into a favourable environment, it is essential that the crop varietal characteristics and its climatic needs and tolerance are intensively analysed in relation to various agricultural and agronomic operations, such as plant protection measures, manure application, weeding etc. which have to be critically timed.

4.5. Breeding Rice Tolerant to Drought with Particular Reference to India

Drought prone areas in India generally receive monsoon rainfall in the range of 700-1000 mm. Rice cultivation in these areas is thus monsoon dependent and gets exposed to moisture stress periodically due to breaks in monsoon. These breaks lasting for periods from 7-10 days mostly in August when the rice plant is actively growing and its water demand is maximum. The drought prone areas constitute about a third of the *kharif* rice area in India and productivity here is highly vulnerable. It is necessary to increase and stabilise rice yield in such lands to keep up the upward trend in rice production as observed in other rice growing areas.

Rice cultivation in areas prone to drought entirely depends on monsoon rainfall and it should be our endeavour to harvest all rain by adopting varieties of suitable duration according to the rainfall pattern.

Four kinds of situations are possible for such a kind of rice cultivation

i. Rains adequate or assured during vegetative phase but adequate during reproductive phase:

ii. Rains inadequate or unreliable during vegetative phase but adequate during reproductive phase;

iii. Rains adequate during vegetative phase but inadequate during reproductive phase; and

iv. Rains inadequate during both vegetative and reproductive phases.

The crux of the problem in rice breeding (exposed to moisture stress) as in items (ii) and (iii), which is often the situation in drought prone area, is to find out suitable donors with drought tolerance during the vegetative and reproductive phases under situation as mentioned in item (iv), rice cultivation is not possible. There is no problem of moisture stress as described in item (i).

Many local varieties suited to such conditions have recently been identified. Through screening, a number of varieties with varying degrees of drought tolerance have been identified e.g. mtu. 17', Mettasannavari' in Punjab and Tkm. 1' variety in drought areas of Tamil Nadu.

Among these, it was found that Lalnakanda - 41' and Ch. 45' have drought tolerance at the vegetative phase, while Mtu. 17' showed drought tolerance even at the reproductive phase.

As most of Tamil Nadu is drought prone where staple food is rice, the first attempt to breed varieties with drought tolerance was made here during the mid-fifties and a drought-tolerant variety "Co. 31" was then released.

With the introduction of Taichung (Native) 1', during 1965, efforts were made to transfer drought tolerance to semi dwarf hybrids, and Bala' from the cross N. 22'xT (N) 1' was the first high-yielding variety with drought tolerance that was released in the country. As Bala' was hard threshing, efforts were made to identify lines with easy threshing and good grain qualities. CR. 113', CR. 115', CR. 141', and CR. 143', had many lines with better threshability and grain qualities than Bala'. One line, viz. 'CR. 141-192'. from the cross (N. 22/T(N) 1xT.90'. IR.8) had been named Kiran in Bihar. Hybrids more productive and tolerant to drought than any of the parents had been identified in the cross CR. 125 (Lalnakanda -41' x Mtu. 17') x T (N) 1.

International Rice Research Institute (IRRI), Manila geneticists have standardised the testing procedure for drought tolerance of rice and have made considerable progress by evolving a large number of promising cultures ideally suited in areas where monsoon often fails. Many of these have been tested in most rice growing countries including India through the International Rice Testing Programme (IRTP). As the problem is faced by every rice-growing state in India, a good number of cultures have been generated by states using local donors, and at present, many are under trial in the AICRIP testing programme.

Efforts were also made at Central Rice Research Institute (CRRI), Cuttack to evolve varieties with drought tolerance through induced mutation in traditional varieties. Considerable success was achieved through this approach and many mutant with higher yield potential and drought tolerance than the parents had been identified in Ch. 45' and Mtu. 17' Of special significance is Mt. 17' Mutant No. 4 which showed very high tolerance to drought even during the flowering phase.

Another mutant from CR. 113' designated CR.M. 13-3241', is possibly the earliest induced productive major cereal in the World, maturing in seventy days when direct seeds under a temperature regime of over 25°C which is often encountered in drought prone areas in India. This mutant yields about 1-1/2-2 tonnes/ha normally but with good management has shown potential up to 5 tonnes/ha. By relying on the earliness of this variety, known or predictable drought spells can be avoided or there is a possibility to raise another rice crop after the drought or flood ravage and the resumption of normal monsoon.

4.6. Stress Conditions and Varieties

As mentioned above, rice is mostly monsoon dependent and hence the crop is often subjected to either low or inadequate rainfall, leading to drought conditions in some areas. In considerable areas (particularly drought prone Punjab and Haryana), salinity or alkalinity also interferes with the normal rice production.

Attempts were made earlier to incorporate drought resistance from wild types e.g. "*O. perennis*", "*O. sativa* var. *fatua*", "*O. sativa* var. *spontania*" and "*O. longistaminata*" into cultivated types. Some selections from the crosses with the local were found to be promising (Chalam, 1959; Richharia and Govindaswami, 1962). Rajagopalan (1957, 1958) carried out systematic work on the breeding of drought tolerant types in the then Madras State (now Tamil Nadu) and a few drought-resistant types and cultures were produced. The physiological aspects of drought resistance relating to rice have been studied at the CRRI, Cuttak since 1962 (Murty *et al.*, 1971). The *indicas*, T (N) 1' and IR. 8' were more drought-resistant than Ponlai *Japonicas*, Tainan-3 and Taichung-65'. IR. 8's were more drought-resistant than Ponlai *Japonicas*.

4.7. Pretreatment and Inducing Drought Tolerance

Various techniques are available to induce drought tolerance in rice through simple agronomic means. Parija and Pillay (1945) recommended the soaking of seeds in water for 24 hours and drying them before sowing to secure better survival of plants after silting. The pre-treatment of seeds at 45°-49°C also induced drought tolerance in the resulting plants (Parija, 1960). The pre-sowing hardening of rice seeds by soaking them in water for 3 days and drying them induced drought resistance. Paricha and Perumal (1974) recommended spraying of 1.5 % CCC at 30 and 45 days after sowing for maintaining favourable water balance during drought at the flowering stage.

Singh and Chatterjee (1980) found that plants grown from pre-treated seeds significantly outscored those from untreated seeds in plant height and population per unit area. Before planting, the treated seeds when sun dried to their original moisture status, were found to increase plant dry matter accumulation ranging from 13 to 54 %.

The increased vigour of the plants from pre-treated seeds were due mainly to the better root growth, particularly of deep roots, chlorophyll, stability index values of leaves.

4.8. References

ANONYMOUS. 1970. A new technology for dry land farming. Indian Agric. Res. Inst. (New Delhi). *Res. Bull. Ser.* **2**: 9-19.

BALAKRISHNA RAO, M.J., SAMPATH, S. and MURTY, K.S. 1971. Breeding varieties for special situations. *Indian Farming*. **21**: 45-52.

BARGER, G.L. and THORN, H.C.S. 1949. A method for characterizing drought index in Iowa. *Agron. J.* **41**: 13-15

CHALAM, G.V. 1959. Botanical improvement of rice with special reference to wild crosses. *Proc. Indian Acad. Sci.* (*B*). **49**: 309-315.

CHOWDHURY, A. and ABHYANKAR, V.P. 1979. Does precipitation pattern foretell Gujarat climate becoming arid? *Mausam*. **30**: 85-90.

DAS, H.P. and DATAR, S.V. 1996. Prospects of double cropping in rainfed rice in West Bengal. *Mausam*, **49**: 121-126

MURTY, K.S. and JANARDHAN, K.V. 1971. Physiological considerations for selection and breeding of rice varieties for saline tracts. *Oryza*. **8(2 suppl.)**: 85-100.

PARIJA, P. and PILLAY, K.P. 1945. Effect of pre-sowing treatment on the drought resistance in rice. *Proc. Natn. Acad. Sci., India*. **15**: 6-14.

PARIJA, P. 1960. Physiological studies on rice. *J. Indian Bot. Soc.* **39**: 335-350.

RAJAGOPALAN, K. 1957. Studies on drought resistance in rice. *Madras Agric. J.* **44**: 194-205.

RAJAGOPALAN, K. 1958. Studies on drought resistance in rice II. *Madras Agric. J.* **45**: 29-54.

RAMAIAH, K. 1953. *Factors affecting rice production. FAO Agr. Dev. paper.* **45**: Rome, Italy, p. 45.

RICHHARIA, R.H. and GOVINDSWAMI, S. 1962. Utilization of the genes of wild rice involving improved cultivated types. *Rice Newsletter*. **10**: 20-21.

SASTRY, P.S.N. 1976. Climate and crop planning with particular reference to rainfall. Proc. Sym. on Climate and Rice, Los Banos.

SINGH, A.I. and CHATTERJEE, B.N. 1980. Effect of seed pre-treatment on rainfed dryland rice production and on water saturation deficit in leaves. *IRRN*. **5(3)**: 19-20.

VAN BAVEL, C.H.M. 1953. A drought criterion and its application in evaluating drought incidence and hazard. *Agron. J.* **45**: 167-172.

YOSHIDA, S. 1975. Major research in upland rice. *International Rice Research Institute Publication*. Los Banos, Philippines.

Rice—in a Variable Climate, 1999, pp. 51-61
Eds. Y.P. Abrol and Sulochana Gadgil

5. Influence of Climate Parameters on Productivity of Rainfed Rice: Case Study for Chattisgarh

In the rainfed rice eco-system, though the seasonal rainfall is mostly greater than 1000 mm., intermittent dry spells, often lead to acute water stress conditions during crop growing season. Also, during monsoon season, solar radiation becomes a limiting factor for better crop's growth and development. According to Islam and Morison (1992), under irrigated conditions in Bangladesh, there was a linear relationship between rice yields and irradiation during reproductive and ripening stages.

In Chattisgarh region of Madhya Pradesh (Fig. 1), rice grown in about 3.8 million hectares of which 80 percent area is under rainfed conditions. The rest of 20 percent area, irrigation is protective in nature and is available after the monsoon season.

In this communication we present an analyis of the effect of the three major regimes viz. the moisture, thermal and light regions on the rice productivity in rainfed rice eco-systems with special reference to Chattisgarh region in Madhya Pradesh.

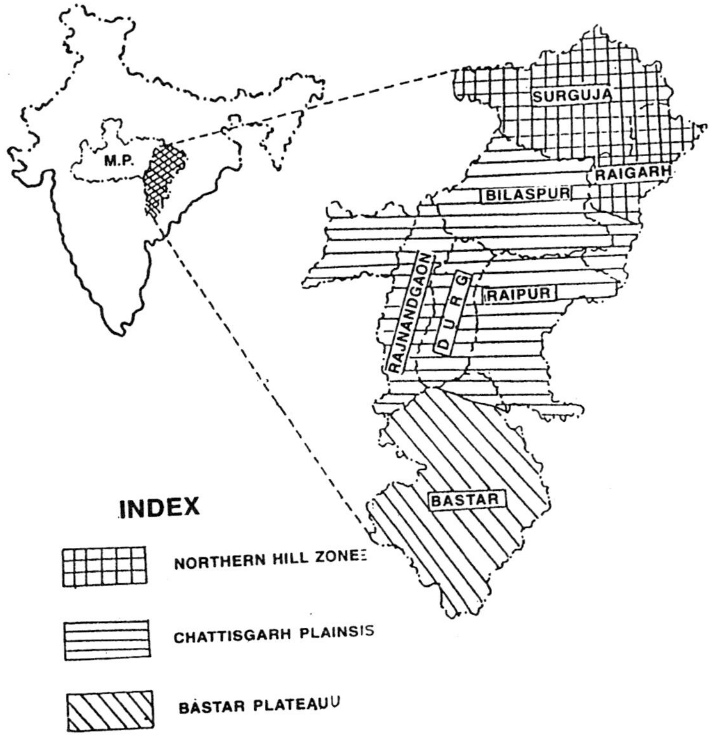

Fig. 1: Location map of Chattisgarh region and different agroclimatic zones

5.1. Ranfall Pattern

5.1.1. Rainfall Quantum

With about 1200-1400 mm of water requirement for medium/late maturing rice varieties, the entire Chattisgarh region can be divided into 4 different zones (Fig. 2) as per rainfall quantum given below:

Rainfall Zone	*Possibility of Rainfed Rice Cultivation*
> 1600 mm	Suitable for rainfed rice cultivation even under transplantation
1400-1600 mm.	Suitable for rainfed rice cultivation under traditional 'broadcast biasi' method or improved line sowing method. Good productivity can be obtained by adopting full package of practices.
1200-1400 mm.	There are possibilities of water stress during crop growth period. Good productivity can be obtained by adopting suitable water conservation and recycling methods and other packages.
Less than 1200 mm.	There are more chances of water stress. This region needs crop diversification slowly. At present, water management practices should be adopted for good crop of rice under rainfed conditions.

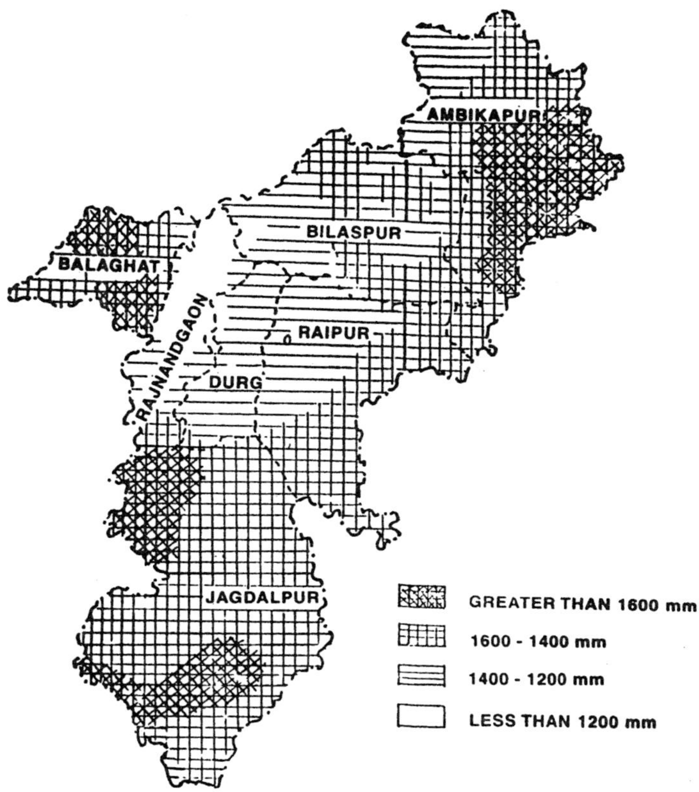

Fig. 2: Annual rainfall in Chattisgarh

5.1.2 Rainfall Distribution

For assessing the suitability of a crop, the rainfall distribution during the crop growth period rather than total quantum of rainfall need to be considered. In view of this, areas suitable for rice cultivation are identified based on the monthly rainfall distribution. The monthly is denoted as:

Category	Monthly rainfall (mm.)
A+	Greater than 400
A	300-400
B	200-300
C	100-200
D	50-100
E	Less than 50

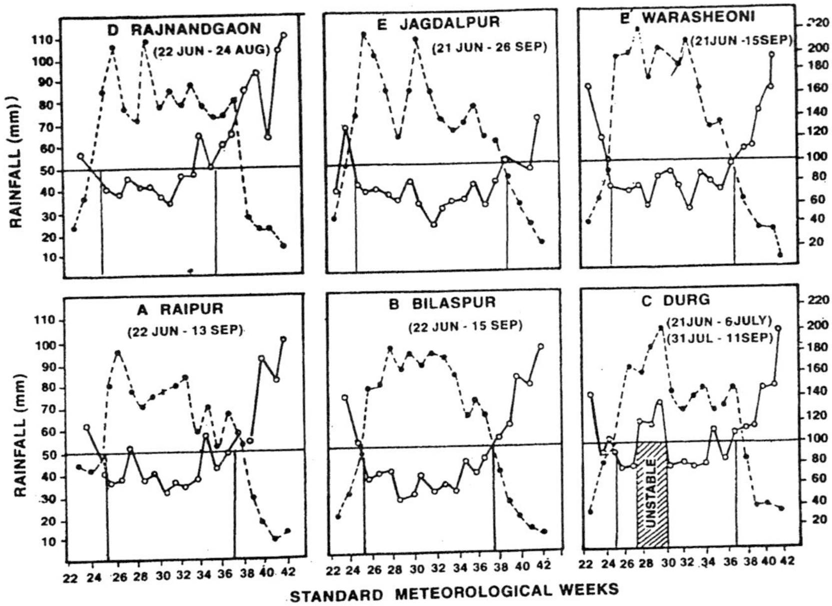

Fig. 3: Stable rainfall periods at different stations in Chattisgarh region

Based on this criterion, the monthly distribution of rainfall is considered to identify areas suitable for rainfed rice cultivation as follows:

$A_2 + B_2$	Highly suitable for rainfed rice cultivation even under transplanted condition.
$A_2 B_2$	Highly suitable for rainfed cultivation of rice under traditional basis system with improved techniques available.
$A_2 B_1 C_1$	Nominally suitable for rainfed rice cultivation with good management and improved techniques.
A_2C_2 or $A_1B_1C_2$	Not economically suitable for rainfed rice cultivation. Crop diversification is necessary in near future.
Other combinations	Not economically suitable for rainfed rice cultivation. Immediate crop diversification is necessary.

Note: Number in subscript denote the number of months under each category.

Using this criterion different blocks in each district under each situation were identified. For example, in Balaghat district, where rainfed rice cultivation under transplanted conditions is being practised and under this classification, almost all the blocks in Balaghat district are coming under A_2^{+} $_2B_2$. In Raigarh district too,

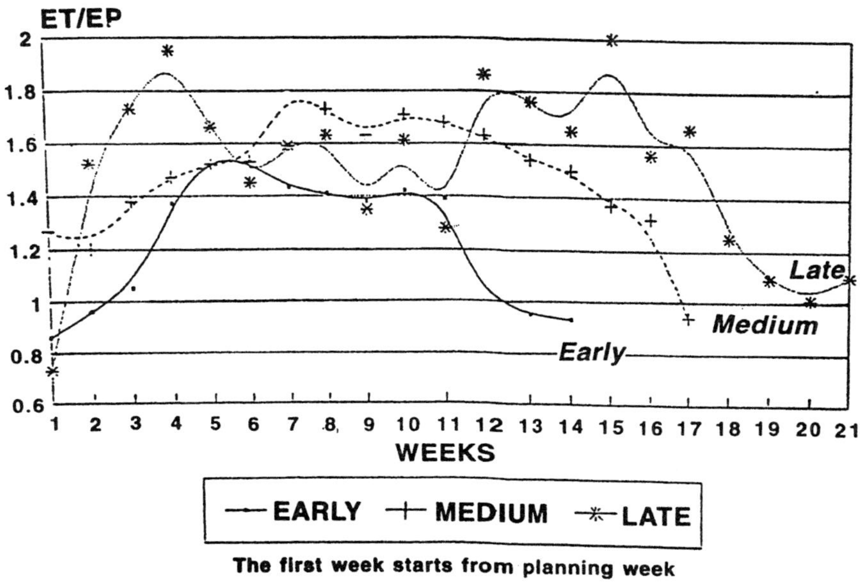

Fig. 4: Pattern of crop coefficients of rice genotypes of three different durations under Raipur condition

all blocks are coming under $A_2^{\ddagger}$ $_2B_2$. (highly suitable) category and rainfed transplantation can be adopted in this district too where it is not being practised now. The areas in each district that come under different categories of rainfed cultivation are shown in Table 1.

5.1.3 Stable Rainfall Periods

For examining the suitability of different rice genotypes of different durations, stable rainfall periods are considered which can be defined as "the periods when the weekly average rainfall is greater than 50 mm and corresponding coefficient of variation is less than 100 per cent". In this region, the daily ET and percolation losses account for 3-4 mm each and, therefore, a minimum rainfall of 50 mm/week is needed as stable rainfall quantum. With this criterion, the stable rainfall periods for each district in Chhattisgarh are shown in Figure 3. It can be seen that the stable rainfall period for Raipur is from 22 June to 13 September while for Bilaspur it is from 22 June to 15 September. However, for Durg the stable rainfall period is coming in two parts i.e., 21 June to 6 July and 31 July to 11 September with instability (higher CV values) in between. In other words, there are chances of instability in rainfall during the seedling stages of rice crop. This suggests the need for a variety with drought tolerance during seedling stage for Durg. Similarly the stable rainfall period at Rajnandgaon is of shorter duration (22 June

Table 1. Areas under different categories of rainfed rice cultivation in all the districts of Chattisgarh region of M.P.

District	Highly successful		Successful	Failure	Most Failure others
	Transplanted A2	Broad cast A2 B2	A2 B1 C1	A2 C2 or A1 B1 C2	
Durg	Nil	Sanjari, Ambagarh Chowki, Bhilai, Khapari	Durg	Kharda, Bemetara	Gandhi
Rajnandgaon	Nil	Rajnandgaon, Dongargarh	Adambad	Khairagarh, Kawardha and Khamaria	Chhuikhadan
Raipur	Gariaband, Mahasamund & Pithora	Raipur, Dhamtari, Arang, Rajim, Saraipali, Deobhog, Pilandaraon, Kusrangi, Lakholi, Rudri, & Madamsilli Kendri	Balodabazar and Kanki	Simga and Kurud	Nil
Bilaspur	Katghora, Ponditafa	Pendra	Janjgir, Korba, Sakti, Champa	Bilaspur, Mungeli, Pendraia, Pasan, Khutaghat, Khuria, Sheorinarayan	Nil
Raigarh	Raigarh, Sarngarh, Daramjaigarh & Jashpurnagar	Nil	Nil	Nil	Nil
Balaghat	Balaghat, Baihar, Lanji, Waraseoni, Haramouni, Dhuti, Lalburra, Sarathi, Kattanjihiri	Katangi, Jamunia, Salitake	Dongarhgaon, Palhera, Falga	Nil	Nil
Bastar	Antagarh, Bijapur, Bhanupratappur	Jagdalpur, Kondagaon, Konta, Keshkal, Kanker	Bhopalpatna, Dantewara, Sukma	Nil	Nil
Surguja	Ambikapur, Baikunthpur	Nil	Nil	Janakpur	Nil

Monthly Rainfall: A = > 400 mm, A = 300–400 mm, B = 200–300 mm, C = 100[200 mm, D = 50–100 mm, E = < 50 mm.
Note: (Numbers in subscript indicate number of months in each category)

to 24 August) and this suggests for recommending early maturing variety or varieties with drought resistant at reproductivity stage for Rajnandgaon.

5.1.4 Onset of Monsoon

In rainfed rice eco-system, especially under monsoon climates, the productivity of rice depends mainly on the vagaries of monsoon. The early, normal and late onset of monsoon affect not only the water availability periods but also the quantum of rainfall. As a consequence, the productivity of rice gets affected. For example, when monsoon onset is delayed, the total rainfall quantum decreases for 1366 mm from normal to 1044 mm and the productivity is reduced from 1.15 t/ha to 0.84 t/ha.

However, when monsoon comes early, the rainfall quantum increases from 1366 mm to 1607 mm, but there is hardly any increase in the rice productivity (Table 2). Hence, there is a need for varietal manipulation depending upon the monsoon onset in the rainfed rice ecosystems.

Table 2. Average rainfall, water availability periods and corresponding average yields of rice during early, normal and late onset of monsoon in Raipur district.

Monsoon onset	Average rainfall (mm)	Water availability periods (days) Moist I	Humid	Moist II	Total	Yield (t/ha)
Early	1607	7	127	12	146	1.18
Normal	1366	14	115	10	139	1.15
Late	1044	13	91	14	118	0.84

Moist I : PE > R > PE/2 Before humid period

Moist II : PE > R > PE/2 After humid period

Humid : R > PE

Where, R = Rainfall, PE = Potential Evapotranspiration

5.2. Evapotranspiration and Crop Coefficients

Based on 14 year lysimetric experimental results, the evapotranspiration (ET) during different growth stages of early, medium and late maturing rice genotypes under Raipur conditions have been worked out and are shown in Table 3. The total evapotranspiration losses during the entire growing season of early, medium and late maturing varieties under Raipur conditions are 461, 627 and 750 mm, respectively. The weekly pattern of crop coefficients (ET/EP) during the growing season of early, medium and late maturing varieties are shown in Fig 4.

Table 3. Evapotranspiration during different growth stages of early, medium and late maturing rice genotypes.

Crop Stage	Early		Medium		Late	
	Duration (Weeks)	ET (mm)	Duration (Weeks)	ET (mm)	Duration (Weeks)	ET (mm)
Seedling (establishment)	2	43.9	3	90.5	3	100.3
Vegetative	5	178.9	6	249.3	8	300.8
Reproductive	4	164.3	6	209.4	6	248.2
Maturity	3	73.8	3	77.7	4	100.7
Total	14	460.9	17	629.9	21	750.0

5.3. Effect of Temperature and Sunshine on Productivity

Experiments were conducted to examine the influence of thermal and light regimes on rice productivity during the period 1992-1993 to 1994-1995 with 24 rice genotypes. Out of which, IR 36 is popular not only in these parts and of India but also in other countries too. Therefore, an analysis has been made on the effect of temperature and sunshine on the productivity of IR 36.

The growing period averages of maximum and minimum temperatures, total number of sunshine hours and corresponding rice productivity of IR-36 in 8 different dates of sowing during the *kharif* and *rabi* seasons of 1992-93 and 1993-94 are shown in Table 4.

Table 4. Growing period averages of maximum and minimum temperatures, total bright hours of sunshine and corresponding grain yield of rice (variety: IR 36) during different dates of sowing.

Date of Sowing	Average Temp. (°C)		Total Bright Hours of Sunshine	Yield Kg/ha
	Max.	Min.		
Kharif:				
20.06.92	31.4	22.5	544.0	3650.0
05.07.92	31.0	22.1	642.8	2050.0
25.06.93	30.8	24.7	558.8	4418.0
10.07.93	30.5	23.9	642.1	3417.0
04.08.93	30.1	19.7	774.0	2200.0
Rabi:				
09.12.92	32.3	17.4	1373.8	5485.0
02.01.93	36.1	18.4	1541.6	4800.0
25.01.93	37.1	19.7	1351.2	6514.0

It can be seen that the crop grown during monsoon season was exposed to higher minimum and lower maximum temperatures while in *rabi* season the reverse was the case. Similarly the crops grown during *rabi* season received 2-5 times higher bright hours of sunshine as compared to monsoon season and the rice productivity is higher in *rabi* as compared to *kharif* season. Thus, it can be inferred that sunshine is a limiting factor for higher rice productivity during monsoon (*kharif*) season.

Using the data base for the 8 dates of sowing, the highest average, lowest average and mean average of maximum and minimum temperature and highest, lowest and mean total of bright sunshine hours for different crop stages as well as for the entire crop growth period are worked out and are shown in Table 5.

Table 5. Range of temperatures, growing degree days and total number of sunshine hours in each growth stage of rice crop. (cv IR-36)

State of Crop	Max. Temp. (°C)	Min Temp. (°C)	Cumulative heat units	Total bright sunshine (hours)	Yield (Kg/ha)
		Initial Seedling Stage			
(i) Highest	34.7	23.6	546.3	487.5	—
(ii) Lowest	28.3	9.8	410.0	70.5	—
(iii) Mean	31.1	18.7	465.6	227.1	—
		Vegetative			
(i) Highest	36.9	23.4	914.7	408.0	—
(ii) Lowest	29.4	19.1	557.5	103.4	—
(iii) Mean	31.4	22.5	675.3	267.4	—
		Reproductive Stage			
(i) Highest	43.0	23.4	915.7	408.0	—
(ii) Lowest	29.5	19.1	558.5	103.4	—
(iii) Mean	34.3	22.5	675.3	267.4	—
		Maturity Stage			
(i) Highest	43.6	26.7	862.3	357.3	—
(ii) Lowest	29.5	12.2	325.7	215.1	—
(iii) Mean	34.7	21.9	601.6	262.7	—
		Whole Crop Growth Period			
(i) Highest	37.1	24.7	2835.8	1541.6	6514.0
(ii) Lowest	30.1	17.4	1789.2	544.9	2050.0
(iii) Mean	32.4	21.1	2222.6	928.0	4067.5

It can be seen that the highest averages of maximum temperature occur during reproductive and maturity stages as during these stages *rabi* crop come under summer season. Similarly, the lowest averages of minimum temperatures come during seedling and vegetative stages for the *rabi* crop. However, the effect of increased yield due to increased total numbers of bright sunshine hours can easily be seen in the averages for entire crop season.

5.4. Effect of Thermal Regime on Genotypes

The effect of thermal regime is different for different genotypes of rice. Based on the experimental work conducted during the 3 different *kharif* and *rabi* seasons with 24 rice genotypes with 3 different sowing dates in each season the effect of thermal regime on different rice genotypes was examined. For this purpose, a thermal sensitivity index (TSI) was developed which is:

$$\text{TSI} = \frac{\text{Range of duration}}{\text{Average duration}} \times 100$$

During monsoon season, the temperature variation is not much and hence the duration of crop/variety does not vary but during *rabi* season the variation in temperatures is very high and hence the durations of crop varieties also vary considerably. The variation in duration of thermally sensitive varieties is higher than that of non-sensitive varieties. Based on this fact, the TSI has been developed.

With the help of TSI, the thermal sensitivity of different genotypes was arbitrarily classified as follows:

TSI (%)	Thermal Sensitivity
0—10	Insensitive
11—20	Moderately sensitive
21—30	Highly sensitive
31 or more	Very highly sensitive

With this classification, the rice genotypes that come under different classification are identified as follows:

Termal Sensitivity	*Genotypes*
Insensitive	Annada, IR 42342-40-3-3-2-2
Moderately sensitive	Tulsi, IR-36, Vikas, IR-64, Abhaya, Surekha, Salivahana, IR-422-35, IR-363-1-2-3
Highly sensitive	Kalinga-3, Ratna, R 370-1, Ruchi, R 370-2.
Very highly sensitive	Vanaprabha, R-281-8832-1, Madhuri, Kranti R 371-1, R 320-300, Jaya and R 321-7

Thus, in Chattisgarh region, all the rice genotypes are not suitable to be grown in both the seasons. Only those varieties which are either insensitive or moderately sensitive to temperatures can be grown in both *kharif* and *rabi* seasons. In other words, out of 24 rice genotypes, only 10 genotypes are suitable to grow during *rabi* season in this area.

5.5. Summary and Conclusions

In the rainfed rice-ecosystems, the moisture regime plays a dominant role as compared to thermal and light regimes. In Chattisgarh regions of Madhya Pradesh, which represents typical rainfed rice-ecosystem, the rainfall quantum varies from less than 1200 mm to greater than 1600 mm. Based on the rainfall quantum, this region is divided into 4 different rice growing environments. Using the monthly rainfall distributions also, different areas under different categories of rainfed rice cultivation have been identified. There are some areas where rice can be cultivated even under transplanted conditions and there are some regions where rice is not a suitable crop and crop diversification is needed.

The onset of monsoon also plays a vital role on the rainfall quantum and water availability periods. But it was found that the productivity of rice is reduced by 20-22 per cent in case of late onset of monsoon, but there is no increase in productivity in case of early onset. Varietal manipulation is needed for exploiting the suitable moisture prevailed during early, normal and late onset of monsoon.

Regarding the light regime, it was found that the productivity of rice is higher in *rabi* season as compared to *kharif* season mainly due to increased bright hours of sunshine. Thus, during monsoon season, light (radiation) is a limiting factor for productivity as cloudy conditions prevail even during drought periods. For assessing the effect of thermal regime on rice genotypes, an index known as thermal sensitive index (TSI) has been developed. Based on TSI, a criterion has arbitrarily been developed to classify different rice genotypes as thermally insensitive, moderately, highly and very highly sensitive. Of the 24 varieties examined, Annada, a popular early maturing variety, was found to be thermally insensitive. The other prominent varieties which are moderately sensitive and can be grown in both *rabi* and *kharif* season are Tulsi, IR-36, Vikas, IR-64, Abhaya, Surekha, Salivahana. The rest of the varieties are not suitable to be grown in *rabi* season.

Rice—in a Variable Climate, 1999, pp. 63-105
Eds. Y.P. Abrol and Sulochana Gadgil

6. Rainfall Prediction for Rice Growing Areas—Potentials and Prospects

The crucial role of climate in agriculture has been recognized from pre-historic times in India. *Rigveda*, *Atharva-Veda*, Aryans (before 3000 BC) and Buddhist time references, '*Arthashastra*' of Kautilya, '*Meghdoot*' of Kalidas have discussed various aspects of climate in relation to agriculture such as cultural practices of agriculture, crop production, soil moisture, rain hazards, for agriculture even weather forecasting, has been attempted by *Varahamihira* in his 125 *slokas* of *Brihatsamhita*. Historical documents about climatic calamities and its link with agriculture and economics are replete viz. Walford (1878), Loveday (1914), Indian Famine Commission (1880, 1898, 1901). Scientific study of impact of climate in agriculture, however, began after availability of meteorological instrumental observations recorded chronologically at different places which have provided a quantitative basis for studying climate and its relationship with other parameters. The attempts to understand the climate-agriculture relationship over the World by workers from disciplines such as micro- and macro-climatology, meteorology, ecology and geography have resulted in the emergence of agro-climatology as a separate branch after 1950 which is broadly 'The study of aspects of climate that are relevant to agricultural problems'. Pioneering work in this direction has been done by L.A. Ramdas from 1922 with special emphasis on various aspects of micro-climate such as soil temperature, soil moisture, air layers near the ground, natural and artificial modification of micro-climate (Ramdas, 1922, 1946, 1951, 1957).

For India, effective long range rainfall forecast is a socio-economic compulsion. The history of long range forecasting research (LRF) in the country is spread over more than 100 years. Blanford (1884) for the first time made an attempt to foreshadow the summer monsoon rainfall (June to September) activities over India using snow cover data from the northern hilly districts of the country. Sir Gilbert Walker (1910, 1914 & 1924) made very valuable and landmark contribution in this endeavour and set out a tradition of modern scientific research of LRF.

So far considerable success has been achieved in long-range forecasting of the all-India summer monsoon rainfall. However, methods for prediction for smaller spatial scales are yet to be developed. To divide the country into smaller regions for climate variability studies and long-range forecasting has been attempted since Blanford (1886). Walker (1914) divided the country into three areas, Shukla (1987) into four regions, Gregory (1989) into ten regions, Gadgil *et. al.* (1993) into 31 regions for SW monsoon studies. The long-range forecasting has been attempted mostly for country as a whole or the three regions viz. North West India, North East India and Peninsular India. Sontakke and Singh (1996) have delineated the country into six optimum regions after examining its spatial patterns and longest precipitation series have been constructed for seasonal, monsoon monthly and annual period (Singh and Sontakke, 1996a, 1999a). For agricultural purposes, however, still smaller spatial and temporal scale i.e. at least district level monthly precipitation studies are needed.

In the first part of this paper (Section 6.1) the variability of weekly and monthly precipitation during the summer monsoon (June to September) and the potential for prediction on monthly scales are analysed for five stations in the rice growing areas of the country. A substantive fraction of the rainfall in the rice-growing regions of the country occurs outside the summer monsoon. The problem of generating predictions for the post-monsoon or the north east monsoon season is addressed in Sections 6.2-6.4.

6.1. Monthly SW Monsoon Precipitation Variability and Potential for Prediction over rice Growing Indian stations

The objective is to examine monsoon monthly precipitation characteristics and variability at the stations from rice growing areas and to investigate potential for developing quantitative long-range prediction schemes. The selected stations are Bahraich (East Uttar Pradesh), Malda (Sub-Himalayan West Bengal), Tezpur (North Assam), Raipur (East Madhya Pradesh) and Puri (Orissa). Locations of these stations are shown in Fig. 7-A. The area in '000 hectares and production in '000 tonnes of rice crops over the districts to which these stations belong according to 1982-83 figures (Muthiah and Ramachandran, 1990) are Bahraich: 218(123), Malda: 216(261), Tezpur: 470(471), Raipur: 371(270), Puri: 392(261), respectively. In this article :

i) Weekly total mean precipitation and unconditional probability are presented for the five selected stations.
ii) The longest updated continuous monsoon monthly precipitation series for Bahraich (1867-1995), Malda (1861-1995), Tezpur (1860-1995), Raipur (1863-1995) and Puri (1848-1995) have been analyzed and presented.
iii) The climatic noise of monthly monsoon precipitation estimated over India and surrounding areas is depicted.
iv) Potential for monthly monsoon precipitation prediction estimated at the selected stations is discussed. In addition, weekly total mean precipitation and unconditional probability have been presented for 5 selected stations useful for rice crop planning.

6.1.1. Data

Summer monsoon monthly precipitation data used for the five selected stations from the earliest available year to 1900 has been obtained from Eliot (1902). From 1901 to 1995 it has been collected from relevant records of Indian Meteorological Department, Pune.

Data used only account for about 2 % of the total have been filled by the ratio method as suggested by Rainbird (1967). All the monthly data sets have been tested for homogeneity by Swed and Eisenhert run test and time series plots are examined for abrupt change or jump, if any both of which show that they are homogeneous.

For estimating climate noise, daily precipitation data of 1569 well spread stations and 60 stations from surrounding countries was used. Indian stations data spanned from 1901-1970 for all the stations and data available from 1975-1984 for 511 stations out of total 1596 was appended. This data obtained from I.M.D., Pune and Climate Prediction Centre (CPC), Washington, DC was archived from NCAR data centre. The source of daily data from 1979-84/93 for 60 stations from the surrounding countries and 37 Indian stations is Global Telecommunication System. The details of these data sets are available in Shea and Sontakke (1995).

6.1.2. Weekly Precipitation

Weekly precipitation amount and its assurance during rainy season are quite important for crops like rice which requires moisture and water throughout the growing period. The weekly total mean precipitation and unconditional probability (Shea and Sontakke, 1995) plots have been presented for the five selected stations (Fig. 1) and its features during rainy season have been discussed here. The period for calculation of weekly total mean precipitation and latitude, longitude are shown in Fig. 1 against each station. The rainy season has been identified as the continuous weeks during which total mean precipitation is at least 20 mm.

(1) Bahraich: Bahraich being in the middle northernmost northern India the rainy season starts late during 24th week (11-17 June). Highest precipitation occurs in the 29th week (16-22 July) of the order of 90.5 mm. During 33rd week (13-19 August) relative minima is noticed which is due to 'climatological singularity' according to Ananthakrishnan and Pathan (1991). The season ends in 40th week (1-7 October). The probabilities vary in the range 14 to 45%.

(2) Malda: Rainy season at Malda starts from 20th week (14-20 May) when pre-monsoon thunderstorms are active over the area. Bay branch of summer monsoon strikes this area in the first week of June. From 24th to 40th week (11-17 June to 1-7 October) mean total precipitation is more than 50 mm and probability exceeds 40 % from 25th week to 37th week (18-24 June to 10-16 September). Season ends by 42nd week (15-21 October). Relative minima though not prominent is noticed in the 31st week (30 July to 5 August).

(3) Tezpur: Tezpur is located in the hilly region of northeast India. Rainy season starts from 14th week (2-8 May) and ends in the

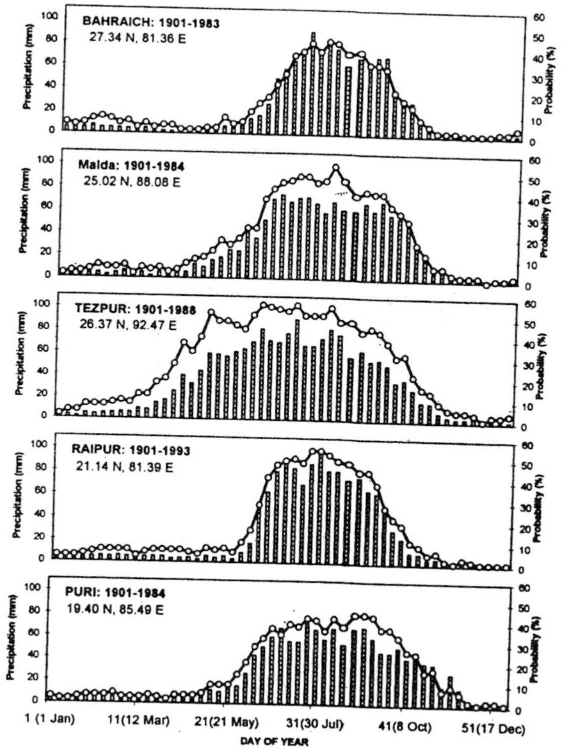

Fig. 1: Weekly total mean precipitation (bar) and unconditional probability of precipitation (line) for the five selected stations based on the period as shown against each station name. The 'x' axis shows week number and starting day of the week. Latitude and longitude of the stations are also given.

41st week (8-14 October). Precipitation totals are higher than 50 mm from 18th to 38th week (30 April-6 May to 17-23 September). During this period probability is at least 40 % and exceeds 50 % frequently.

(4) Raipur: Raipur comes under the influential area of monsoon depressions. From the beginning of the rainy season i.e. 24th week (11-17 June) upto 37th week (10-16 September) rainfall amount exceeds 50 mm. Season ends in the 40th week (1-7 October). Relative minima is observed during 29th week (16-22 July). Probability exceeds 40 % from 25th to 36th week.

(5) Puri: Rainy season starts from 23rd week (4-10 June). Being on the upper east coast through weak North East monsoon is active over Puri from October. Rainy season ends therefore late in 44th week (29 October-4 November). Rainfall total and probabilities are not very high during the whole season.

6.1.3 Monthly Precipitation Variability

Statistical characteristics of the precipitation series viz. mean, median and year to year variability for the summer monsoon months i.e. June, July, August and September are computed based on the common available period 1871-1995 for the selected five stations. Also the series have been examined for frequency distribution by chi-square statistic, for persistency by auto-correlation, and for long term trend by Mann-Kendall rank statistic. The results have been presented in Table 1. The correlation coefficients between precipitation series from April to September at individual stations have been computed but all were found insignificant and, therefore, are not tabulated. Time series of percentage departure from normal i.e. $\{[(x_i - \bar{x})/\bar{x}] * 100\}$ (bars) are plotted in Fig. 2 to Fig. 6 alongwith 9-pt gaussian low pass filter values (curve) for examining fluctuations in the low-frequency mode. The variance explained by filtered series is also given.

(1) Bahraich (1867-1995): Only July series is found gaussian distributed. September series shows persistence significant at 1% level. However, none of the series are showing long term trend. Year to year variability is large for June and September series. June precipitation has experienced high positive departures frequently from 1970 onwards. From 1944 onwards August precipitation is having mostly negative departures showing downward tendency. High positive departures in September series are clustered during 1911-19 and 1980-85.

(2) Malda (1861-1995): June and July series are gaussian distributed. None of the series are persistent. June series shows downward trend significant at 5% level. Year to year variability is not very large. September series is having more negative departures from 1949 onwards.

(3) Tezpur (1860-1995): All the four series are gaussian distributed and free from persistence. August series shows significant negative trend at 5 % level. Year to year variability is small for all the series.

(4) Raipur (1863-1995): Except July other series are non-gaussian. June series is persistent and shows downward trend significant at 5 % level. Year to year variability is large in June and September series. Downward tendency is evident from filtered series in June and September series in recent three decades.

(5) Puri (1848-1995): Only July series is gaussian distributed. Long term trend is not found in any of the series but in recent two three decades all the series are showing upward tendency. Year to year variability is large in June series.

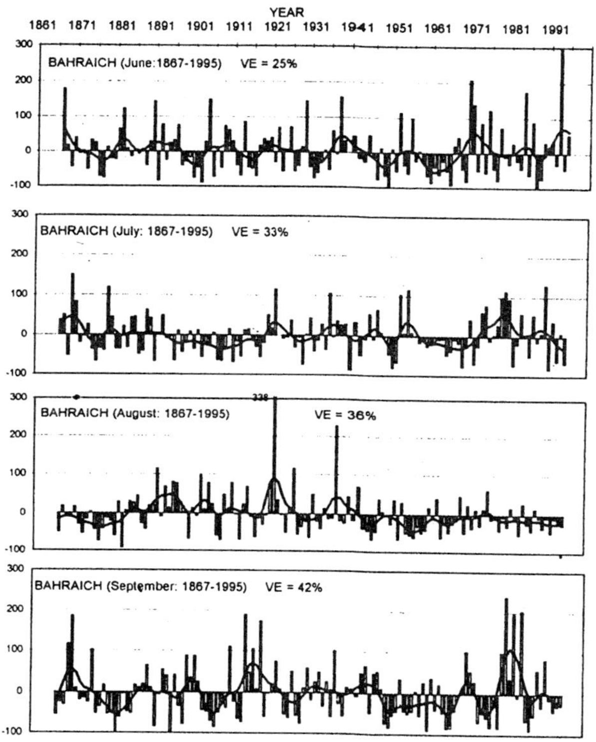

Fig. 2: Percentage departure from normal (bar) and 9-pt gaussian low-pass filtered (curve) series for four monsoon months at Bahraich for the period 1867-1995. "VE" indicates the percentage of variance explained by filtered series of actual series.

6.1.4. Potential for Monthly Precipitation Prediction

Monthly or seasonal value of climatic parameter is obtained by summing/averaging daily values. Daily values varies within a month or a season and monthly or seasonal values varies from year to year. Variance due to change in monthly mean value from year to year is called 'Climate signal'. This change arises due to changes in slow external systems which includes climatic components as well as external influences such as volcanic eruptions, sun-earth geometry. The climate signal is supposed to be potentially predictable. Daily variance which arises due to internal dynamics of the atmosphere is called 'Climate noise' and exists even in unchanging climate. Theoretically, daily variance can be predicted at the most one or two weeks in advance. Daily variance thus is obstreperous for the climate prediction as it is not predictable at climatic scale and is, therefore, called as climate noise (Leith, 1978). Inter-annual variance of a month or a season is the combined measure of 'climate signal' and 'climate

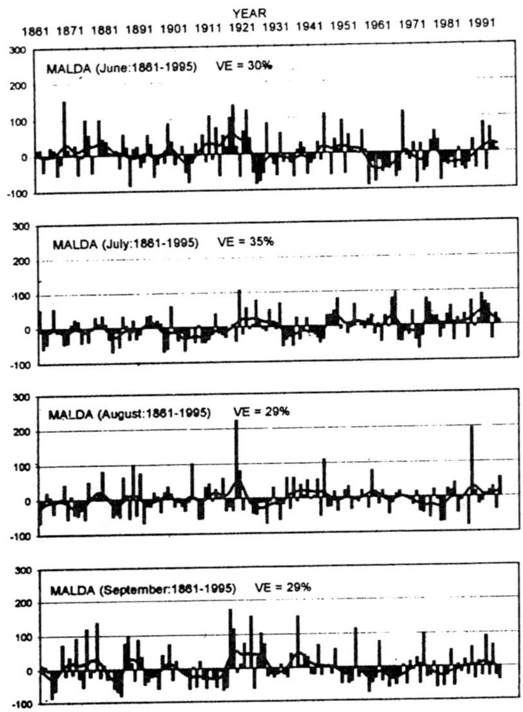

Fig. 3: Percentage departure from normal (bar) and 9-pt gaussian low-pass filtered (curve) series for four monsoon months at Malda for the period 1861-1995. "VE" indicates the percentage of variance explained by filtered series of actual series.

noise' and these two components are approximately separable (Madden, 1981). The ratio of inter-annual variance to climate noise i.e. F-ratio gives us estimate of climate signal which is a measure of the potential for prediction.

Katz and Perlange (1995) have modelled daily precipitation by a two-state first order markov process. Accordingly climate noise,

$$\sigma_T^2 = T\left[qp\sigma^2 + p(1-p)\frac{1+d}{1-d}u^2\right] \text{-------- } Eqn\ 1$$

where, T = length of month (30/31 days)

p = unconditional probability of wet days,

(wet day is day on which precipitation is > 2.5 mm)

q = auto–correlation within sequence of wet days,

u and σ^2 = mean and variance using wet days only,

d = persistence parameter: p11–p01,

$p11$ = probability of a wet day given the previous day was wet,

$p01$ = probability of a wet day given the previous day was dry.

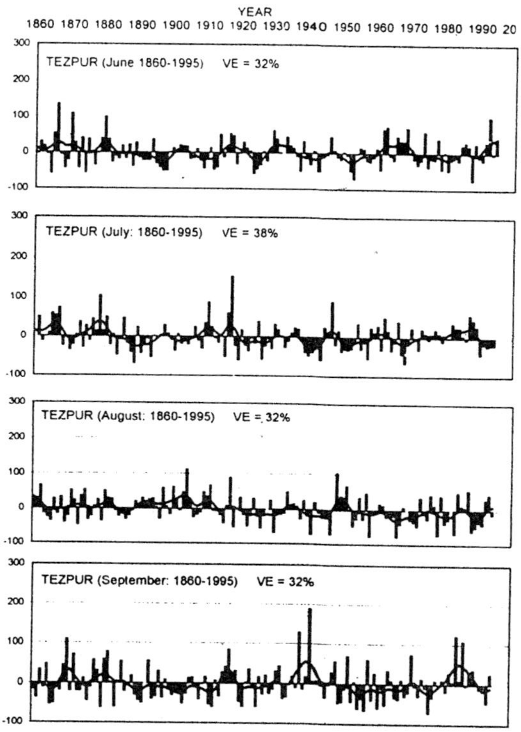

Fig. 4: Percentage departure from normal (bar) and 9-pt gaussian low-pass filtered (curve) series for four monsoon months at Tezpur for the period 1860-1995. "VE" indicates the percentage of variance explained by filtered series of actual series.

Climate noise was estimated by Eqn. 1 for each of the 1656 stations over India and surrounding countries for four monsoon months during each year. Next they were pooled over all the available years. Contours of climatic noise for June, July, August and September are presented in Fig. 7 to Fig. 10, respectively. Unconditional probability of precipitation of wet days (p), average wet day precipitation (u) and persistence parameter (d) are also presented in these figures which depict daily precipitation characteristics. p and u are self explanatory and, therefore, are not discussed separately. Persistence parameter d is high in June over Raipur and Tezpur, in July, over Bahraich, in August near Raipur and in September over all the stations except Malda. The individual p11 and p01 contours show high values near Bay of Bengal and over northeast India in all the months. Persistence parameter i.e. the difference of p11 and p01 is therefore not large over this area. Climate noise is high over northeast India and adjoining parts for all the four months with gradient generally from east to west.

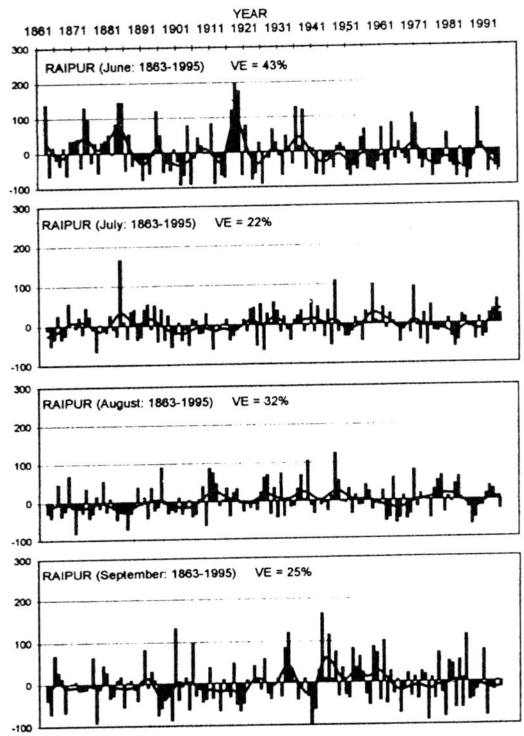

Fig. 5: Percentage departure from normal (bar) and 9-pt gaussian low-pass filtered (curve) series for four monsoon months at Raipur for the period 1863-1995. "VE" indicates the percentage of variance explained by filtered series of actual series.

For each of the four months and also for the season (June-September) the F-ratios and percentage variance expected to be explained out of total are tabulated for each of the stations (Table 1). F-ratios at Bahraich in all the four months are good showing potential to explain about 40 to 43 % variance out of total. At Malda, F-ratio has decreased from June to September with potential of explaining 33 to 17 % variance. At Tezpur, F-ratio for all the months is low. For June and September precipitation it can explain 17 % variance whereas for July and August 38 and 33 %, respectively. At Raipur, F-ratio is higher for June precipitation showing potential to explain 41 % of total variance and lower in August showing potential to explain 23 % variance. F-ratios at Puri in June and July have potential to explain 33 % variance. For August and September they have decreased with potential to explain 23 % and 17 % variance, respectively.

Table 1: Statistical characteristics (1871-1995) and F-ratios of monsoon monthly precipitation at five selected stations

	Mean (mm)	Median (mm)	Yr to yr variability (mm)	Chi-square statistic	Auto-correlation r_1	Mann-Kendall statistic	F-ratio (potential Var. in %)
Bahraich						For season	1.75 (43)
June	150.1	128.1	75.9	20.2***	-.098	-.056	1.75 (43)
July	333.2	309.0	51.5	11.5	.088	.072	1.66 (40
August	295.5	271.0	57.2	23.5**	-.118	-.070	1.70 (41)
September	218.4	189.0	68.2	43.2***	.317***	-.008	1.70 (41)
Malda						For season	1.55 (35)
June	238.2	234.7	58.4	9.7	.081	-.136*	1.80 (44)
July	307.5	295.5	41.3	11.7	.086	.090	1.70 (41)
August	271.4	254.3	51.4	14.3*	-.090	.009	1.60 (38)
September	269.1	224.4	59.8	40.8***	-.025	-.058	1.20 (17)
Tezpur						For season	1.55 (35)
June	302.1	296.3	35.5	5.7	.106	.002	1.20 (17)
July	350.0	337.1	37.1	4.0	.123	.052	1.60 (38)
August	315.9	308.1	41.2	5.8	.062	-.137*	1.50 (33)
September	217.0	209.0	47.9	10.7	.026	.045	1.20 (17)
Raipur						For season	1.50 (33)
June	215.1	182.8	63.6	17.5*	.200*	-.126*	1.70 (41)
July	371.6	354.6	45.5	7.2	-.200	-.048	1.50 (33)
August	356.7	330.8	43.3	16.3*	.024	.056	1.30 (23)
September	210.5	196.3	61.1	23.5**	-.110	.050	1.40 (29)
Puri						For season	1.50 (33)
June	187.3	159.4	82.5	34.0***	-.075	-.018	1.50 (33)
July	269.3	265.1	52.6	7.5	-.021	0.44	1.50 (33)
August	279.6	247.6	50.9	21.5**	-.044	.045	1.30 (23)
September	245.3	224.9	47.6	21.5**	-.049	.029	1.20 (17)

* significant at 5% level ** significant at 1% level *** significant at 0.1% level

Seasonal F-ratios are almost nearby to monthly F-ratios. Monthly F-ratios have even exceeded seasonal F-ratios in some cases. This shows that at station level SW monsoon monthly and seasonal precipitation prediction schemes can be developed with equal scope. It should be noted that F-ratio estimation has following limitations, i) estimation of climate noise depends upon daily precipitation model

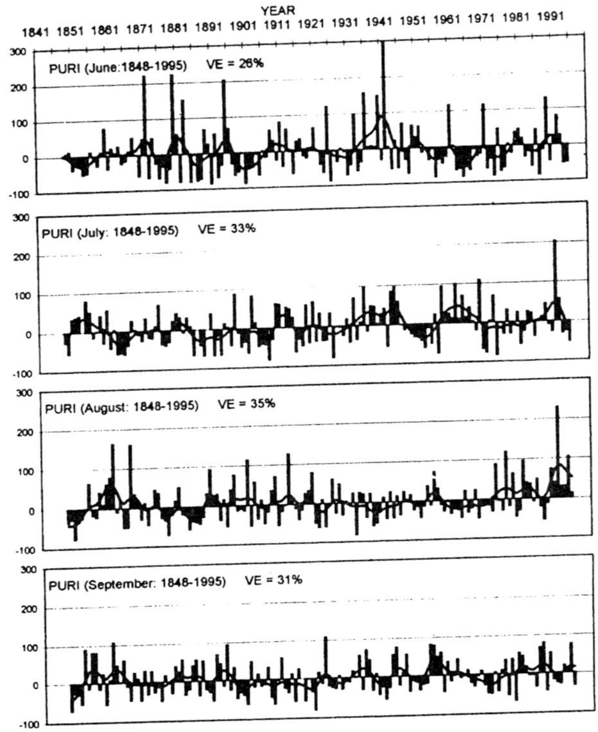

Fig. 6: Percentage departure from normal (bar) and 9-pt gaussian low-pass filtered (curve) series for four monsoon months at Puri for the period 1848-1995. "VE" indicates the percentage of variance explained by filtered series of actual series.

which can be modified further, ii) climate noise and climate signal are not totally independent and, therefore, approximately separable, iii) through sufficiently long for stability particular period has been used in estimating F-ratios. However, F-ratios give good idea about the scope and also limitations for monthly monsoon prediction at station level and should be taken as guideline for developing prediction schemes.

Figs. 2-6 shows the 9-pt gaussian low pass filtered series for monsoon month's precipitation at each station and the variance explained (VE) by it is of actual series. Comparison of filtered series and its VE with variance potentiality can be explained (Table-1) for each station of each monsson month gives broad idea of climate signal in each case. It is felt that the original series should be smoothed by applying filter to the extent that variance amount potentially remains predictable. The prediction schemes should be developed on such smooth series which will be more realistic approach. Also the circulation parameters on regional scale are expected to be more appropriate

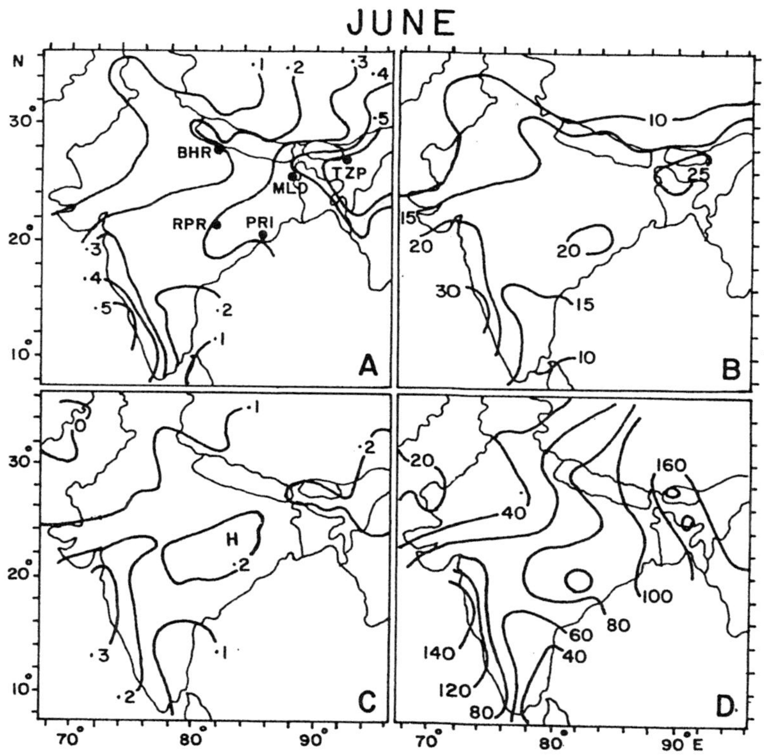

Fig. 7: Unconditional probability of precipitation (A); Average wet day precipitation (B); Persistence parameter (C); and Climate noise (D) of June precipitation over India and its surroundings.

in addition to global circulation parameters for these prediction schemes. Attempts will have to be made to search predictors separately for each station taking into account its location in relation to synoptic systems associated with SW monsoon such as monsoon depressions, monsoon lows, sub-tropical cyclones, upper level westerly waves. Extrapolation of the smoothed series is expected to be a good tool in addition to predictors.

6.1.5. Conclusions

1. Rainy season is longer over Malda, Tezpur and Puri due to early onset of SW monsoon and weak NE monsoon activity. The periods of relative minima in SW monsoon precipitation are noteworthy at Bahraich, Malda and Raipur.

2. The SW monsoon monthly precipitation is not necessarily gaussian except at Tezpur. Mooley (1973) has found that gamma distribution shows good fit to monthly rainfall in the Asian summer

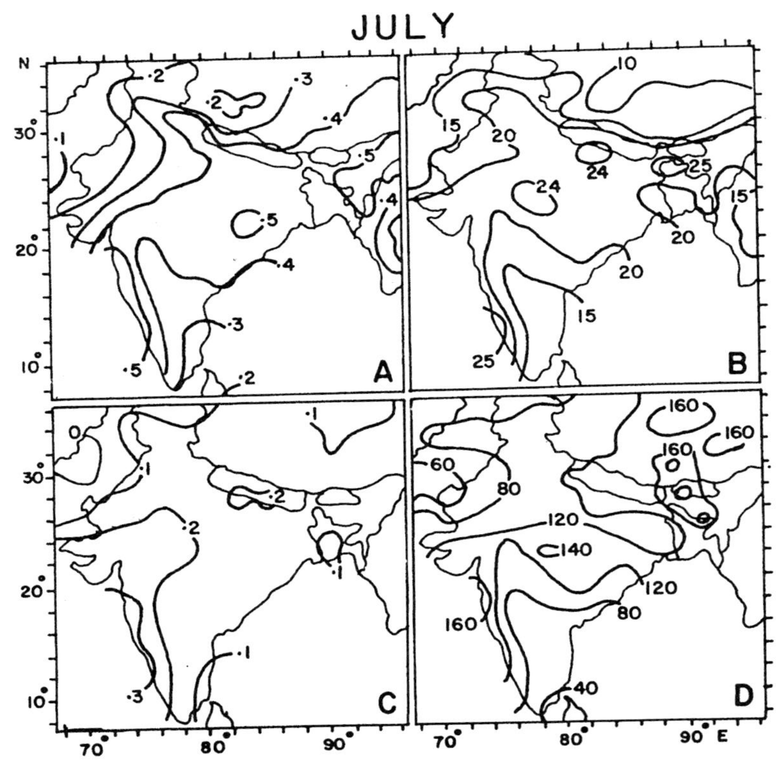

Fig. 8: Unconditional probability of precipitation (A); Average wet day precipitation (B); Persistence parameter (C); and Climate noise (D) of July precipitation over India and its surroundings.

monsoon. Precipitation in June at Malda and Raipur and in August at Tezpur is showing downward trend. Malda experiences negative departures from normal frequently. In latest two to three decades, June and September precipitation at Raipur shows downward tendency whereas Puri precipitation in all the four months shows upward tendency. Each monthly series are found, however, independent where across correlation coefficients (CCs) at individual stations are insignificant.

3. Monsoon monthly and seasonal precipitation is potentially predictable at the selected stations with potential to explain variance in the range 17 to 44 %. This investigation shows scope and limitations for monsoon monthly and seasonal prediction schemes at the stations. Prediction schemes are expected to be more successful on smoothed series by filtering out the climate noise. Search for the predictors will have to be attempted separately for each station where regional parameters may serve better.

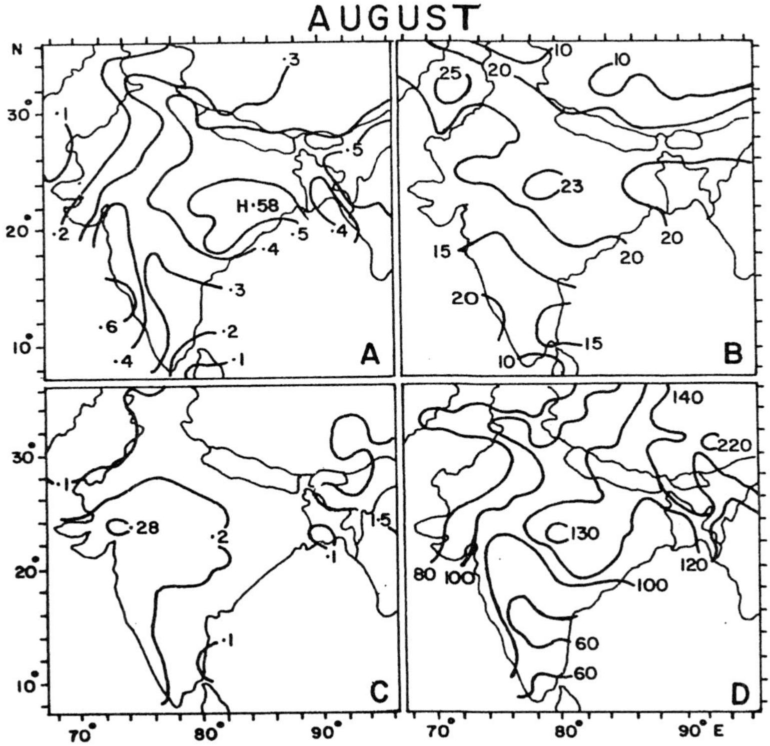

Fig. 9: Unconditional probability of precipitation (A); Average wet day precipitation (B); Persistence parameter (C); and Climate noise (D) of August precipitation over India and its surroundings.

4. Attempt should be made to extrapolate the long series by say curve/surface fitting, trend line; with periodicities; harmonic analysis for future scenario.

6.2. Problems and Prospects of Long-range Rainfall Prediction in Traditional Rice Growing Areas of India

Partly because of well set trend and partly convenience most of the long range rainfall (LRF) studies have attempted to develop predictive model for the total rainfall of the summer monsoon season over the whole country. Owing to large spatial variability even in the total seasonal rainfall, the representation of the all-India summer monsoon rainfall (SMR) series is limited (Singh *et al*, 1991). Therefore, the analysis of the all-India series is of limited practical value. The structure of the isolines of CC between all-India SMR series and the individual stations is not a balanced one. The CC exceeds 0.6 over

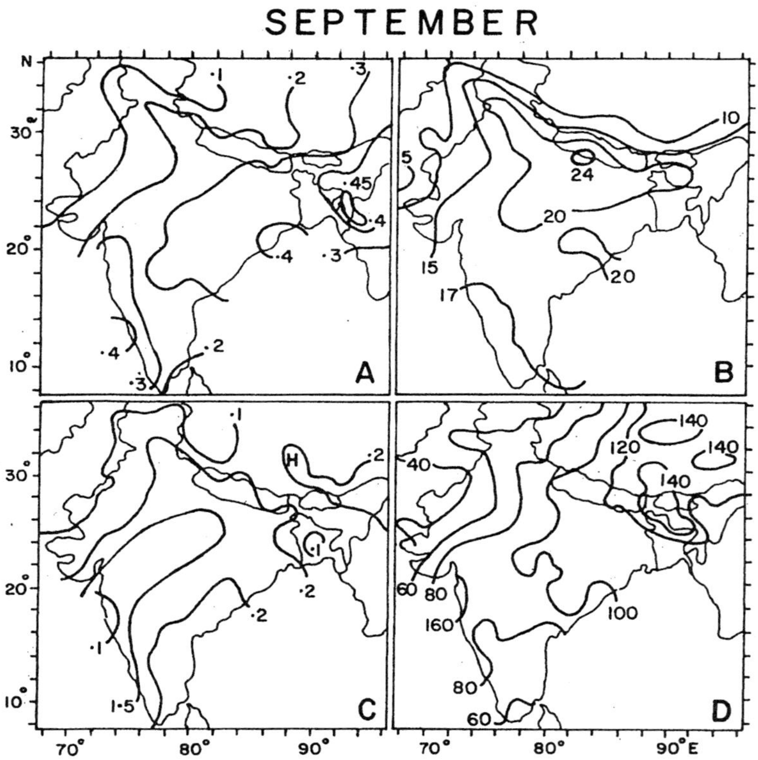

Fig. 10: Unconditional probability of precipitation (A); Average wet day precipitation (B); Persistence parameter (C); and Climate noise (D) of September precipitation over India and its surroundings.

northern parts of coastal Maharashtra. This is the maximum representation provided by the all-India series for any part of the country. The correlation decreases away from this core area and the isoline of 0.1 can be seen over extreme southeast peninsula and over eastern Bihar and northern Orissa. Over northeastern parts, rainfall variations are independent of the all-India mean rainfall variations. It is ironic to see that the representation of all-India series is greater over relatively low rainfall areas and decreases towards the higher side. Therefore, all-India rainfall forecast is not applicable over northeastern, eastern and southern parts i.e. the agriculturally potential areas for rice cultivation. There does not seem to exist any separate LRF research program exclusively for these areas. Due to northeast monsoon rainfall (NEMR) immediately following the summer monsoon they also experience much longer duration of rainy season, about 150 to 200 days (Singh, 1986). Normally, agricultural practices expect two successful consecutive rice crops under rainfed conditions. With the introduction of new high-yielding hybrid varieties, the input cost of rice cultivation, such as seed, fertilizer, pesticide and different agricultural operations, has gone up very high, particularly for upland rice cultivation. For the success and progress of the agricultural enterprise impressive return from the farmers' field is a necessity. Summer monsoon forecasting programs are not seriously concerned with these areas, and northeast monsoon forecasting is not in the agenda of the research scientists. The present initiative is to explore the predictive potential of the two monsoons on regional scale. Besides conventional approach of computing CC between the regional monsoon rainfall series and the circulation parameters representing some aspects of atmospheric phenomena, an alternate new approach is also explored if circulation parameters do not show significant correlation or show weak correlation.

In the following sub section, we briefly describe the procedure for dividing the country into an optimum number of zones for effective studies of SMR variations. Section 6.2.2 gives a brief description of the physical mechanism of the atmospheric circulation features chosen for the study. The SMR and NEMR series considered to examine the association with the circulation parameters are given in Section 6.3. A new time series modelling and extrapolation approach to long range rainfall prediction is developed in Section 6.4 wherein its application to the NEMR prediction over South Peninsular India is also demonstrated. Section 6.5 contains few concluding remarks.

6.2.1. Division of the Country into Homogeneous Zones

Due to vast geographical size and large spatial variability in rainfall occurrences a need was always felt since the inception of modern meteorological research to divide the country into homogeneous rainfall zones. As summer monsoon rainfall of June through September contributes more than 70 % over major parts of the country, spatial coherency in the SMR inter-annual fluctuations is adopted as the criterion for identification of the homogeneous zones (Walker, 1924; Shukla, 1987; Gregory, 1989; Parthasarathy and Yang, 1995; Sontakke and Singh, 1996). The latest approach by Sontakke and Singh (1996) is briefly described here. To identify stations showing similarity in the SMR inter-annual fluctuations the following four types of analyses of the 114 years (1871-1984) of SMR data of 306 stations from relatively plain and contiguous areas of the country (excluding the northern mountainous areas consisting of Jammu and Kashmir, Himachal Pradesh, Hills of Uttar Pradesh and Arunachal Pradesh and the islands in the Bay of Bengal and the Arabian Sea) were carried out. (The study area is 2880049 km^2 which is about 88% of the actual geographical area of the country (3280483 km^2), and this will be referred to as the entire or whole country or all-India).

(*i*) correlation coefficient between all-India SMR and the individual gauges (1871-1980);

(*ii*) change in mean SMR from the period 1901-1940 to 1941-1980 at the individual sites;

(*iii*) the S-mode empirical orthogonal function of the SMR series; and

(*iv*) cluster analysis based on correlation coefficient as well as distance coefficient between the gauges.

Subjective judgmental approaches were applied on the results of the above analyses to carefully choose the geographical meridians and parallels to divide the country into the following six zones (Figure 11).

i. North West India (NWI): north of 21°N and west of 80°E
ii. North Central India (NCI): north of 21°N and between meridians of 80°E and 88°E
iii. North East India (NEI): north of 21°N and east of 88°E
iv. West Peninsular India (WPI): between parallels of 15°N and 21°N and west of 79°E
v. East Peninsular India (EPI): between parallels of 15°N and 21°N and east of 79°E
vi. South Peninsular India (SPI): south of 15°N

The total geographical area of NWI is 34.63% of the study area, of NCI 20.45 %, of NEI 7.95 %, of WPI 14.38 %, of EPI 11.08 % and of SPI 11.60 %. Figure 11 also presents the correlation coefficient (CC) between area-averaged zonal SMR series and its individual gauges. The zonal series is prepared by averaging the gauges in the particular zone. There are 116 raingauge stations in NWI, 65 in NCI, 25 in NEI, 37 in WPI, 22 in EPI and 41 in SPI. The isocorrelate structure showing variation between 0.4 and 0.6 demonstrates a balanced representation of the particular zone by its area-averaged rainfall series, and it is comparable among different zones except EPI in which case it is around 0.6 everywhere. The CC of 0.4 and 0.6 is significant at 0.1% level and better. The correlation structure of this magnitude by the all-India series is shown only over western/northwestern parts. Thus to achieve reasonable representation by the area-averaged series division of the country into a minimum of six zones is essential. The NWI and WPI can be grossly identified as the areas of large-scale dry land cultivation, and the remaining four zones as traditional large-scale rice cultivating areas.

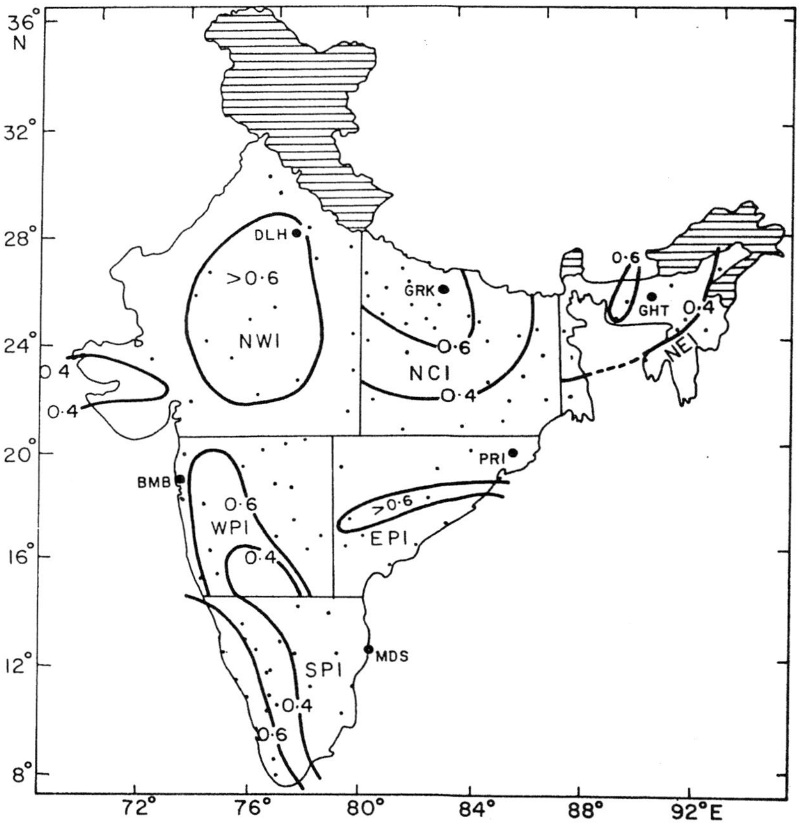

Fig. 11: The six homogeneous rainfall zones of the country. The isoline shows the correlation coefficient between the area-averaged zonal SMR series (1871-1984) and its individual gauges; and the dots are the location of the 116 IMD raingauge stations for routine updating of the different rainfall series of the zone and the all-India.

The six homogeneous zones are in different rainfall regimes. An account of the period of the rainfall occurrences (or rainy season) is given to apprise the different rainfall regimes of the country. The rainy season (or the period of reliable rainfall) can be defined as the continuous period during which coefficient of variability (C.V.) of the monthly rainfall is less than 100% (Singh, 1986). During such months the mean rainfall is more than 50 mm, and the probability of non-occurrence of rainfall is practically negligible (<5%). For individual stations the consecutive months each with mean rainfall more than 50 mm is taken out as the expected rainy season. The starting date is determined in the first month giving a margin of 50 mm from the beginning of the month by linear interpolation. Similarly the cessation date is determined in the last month giving a margin of 50 mm at

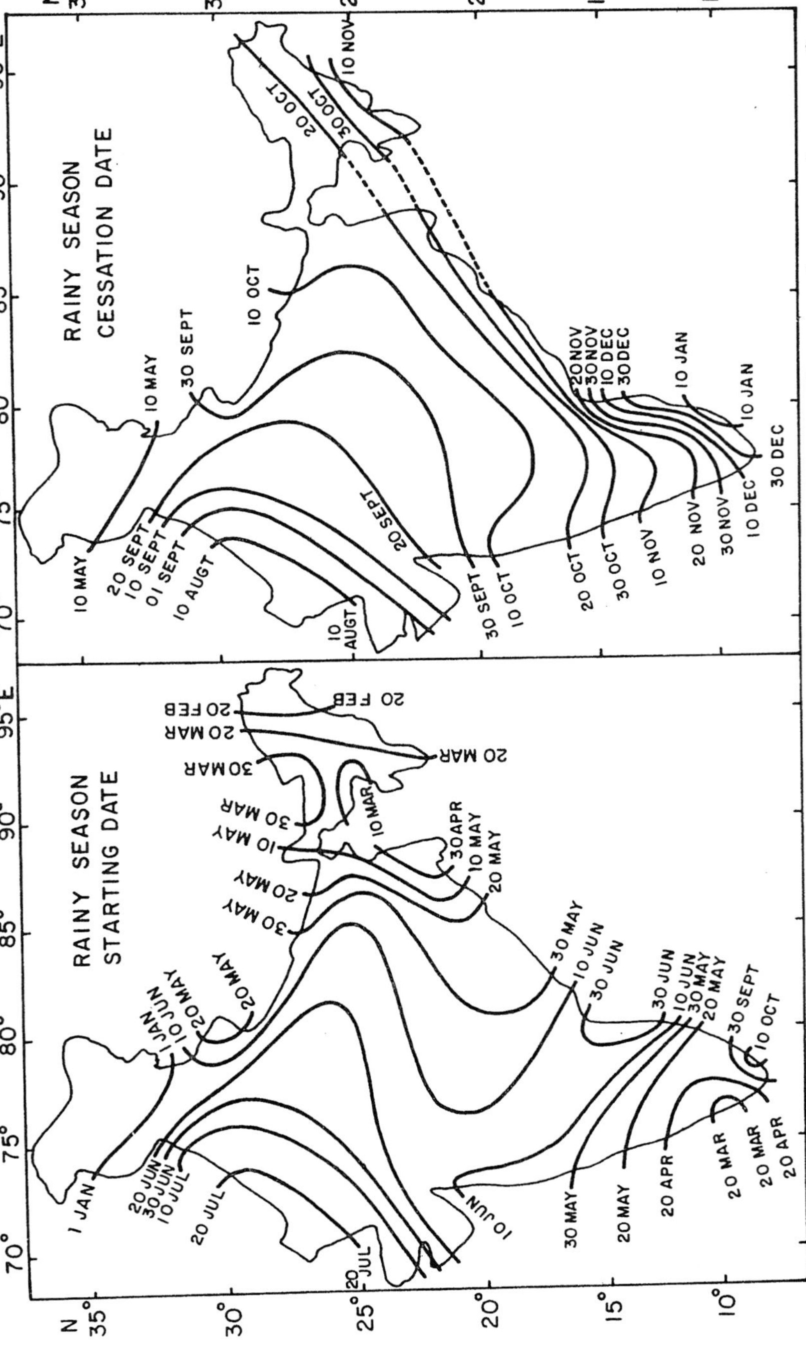

Fig. 12: Normal dates of start (a) and cessation (b) of the rainy season (or the period of reliable rainfall) over different parts of India (after Singh, 1986).

the end of the month. The starting and cessation dates over different parts of the country identified using normal monthly rainfall of 466 stations are presented in figures 12(a) and 12(b) and the duration of the season in figure 13(a).

Over southwest peninsula and the northeastern/eastern India the season starts around 20th March and over northwestern India around 20 July. Over most parts of the country the starting date progressively shifts between these two extremes. The start of rainy season over southwest peninsula and northeast India much before the onset of the southwest monsoon is due to reliable rainfall occurrences associated with thunderstorm activities. The late start over northwestern parts is because of lesser reliable rainfall activities during onset phase of the southwest monsoon over the region. Due to rainshadow effect for the southwest monsoon, the extreme southeast peninsula gets reliable rainfall only during the post-monsoon (or northeast monsoon) season and the date varies between 15 August and 10 October; from north to south. Being the area outside the monsoon regime, Jammu and Kashmir gets rainfall from extratropical systems like western disturbances and trough in the westerlies. Therefore, the season over the region starts between 20 December and 1 January. The cessation date is spatially more coherent, organized and progressively shifting from northwest to southeast. The extreme dates are 10 August for northwestern India and 10 January for southeastern peninsula. For Jammu and Kashmir, the cessation date is 10 May. Consequently, there is a large variation in the duration of the rainy season over different parts of the country, about 20 days over northwestern India to exceeding 240 days over southwestern peninsula. The precipitation intensity during the rainy season also shows large spatial variations from 2 mm/day in northwestern India to exceeding 15 mm/day in west coast (Figure 13a). Based on features of the rainy season, a climatic classification of the country has been suggested for agricultural purposes. Two features namely precipitation intensity in three categories, less than 5 mm/day, between 5 mm/day and 10 mm/day and greater than 10 mm/day, and the duration in three categories, less than 75 days, between 75 days and 150 days and greater than 150 days, are considered in the classification. The six types of rainfall zones identified are as follows (Figure 13b).

Dry Zone-I: The duration of the rainy season is less than 75 days and seasonal precipitation intensity also less than 5 mm/day. Cultivation is possible only during years with above normal rainfall conditions.

Dry Zone-II: The duration is more than 75 days (inclusive of more than 150 days) but the intensity less than 5 mm/day. Under normal conditions one crop can be raised without supplemental irrigation if it is adjusted around the middle of the season.

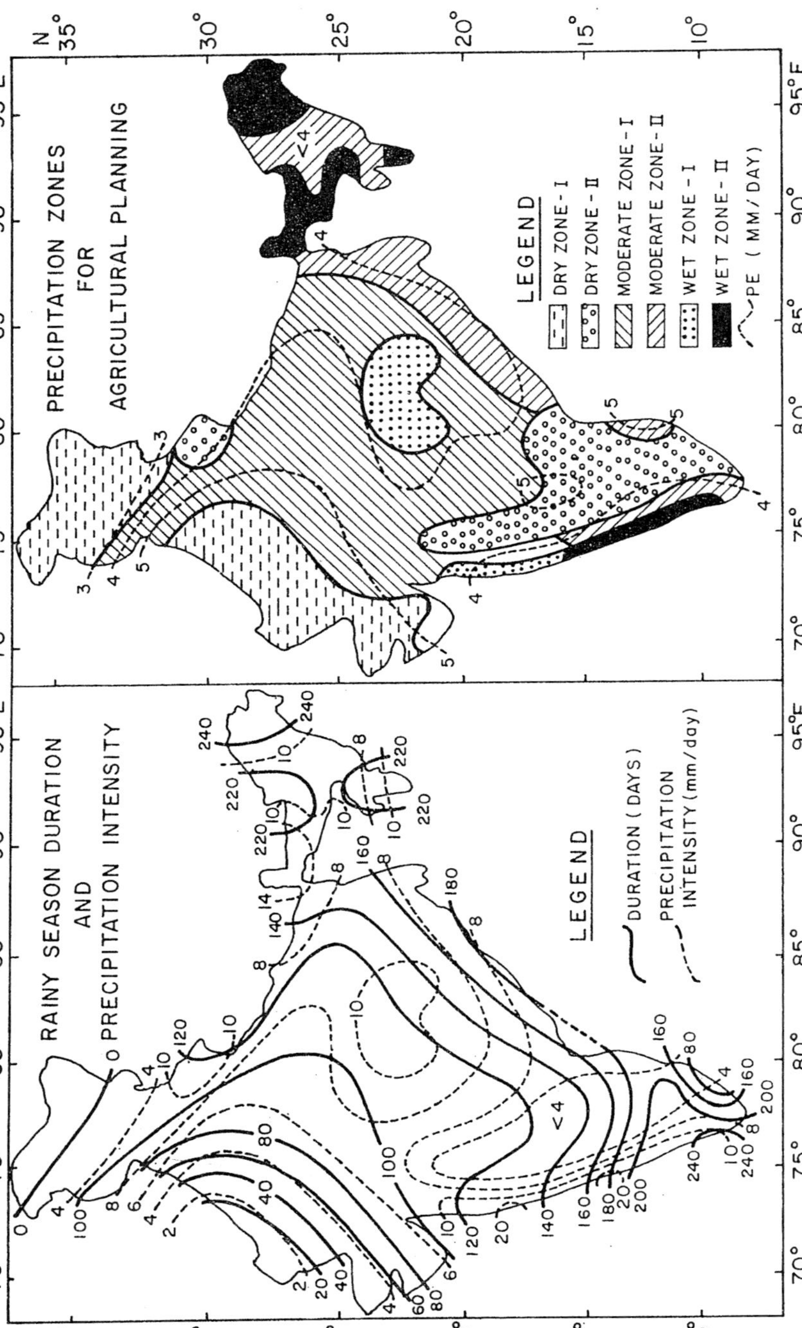

Fig. 13: Normal duration of the rainy season over different parts of India (continuous line) and precipitation intensity during the season (broken line) (a); and a classification of the country for agricultural purposes (b) (after Singh, 1986).

Moderate Zone-I: The duration varies between 75 to 150 days, and the precipitation intensity between 5 and 10 mm/day. Short to long-duration crops can be raised without supplemental irrigation in normal years.

Moderate Zone-II: The precipitation intensity varies between 5 to 10 mm/day but the duration is more than 150 days. Two crops can be raised without supplemental irrigation.

Wet Zone-I: The duration is between 75 to 150 days but precipitation intensity exceeds 10 mm/day. One long-duration and high water requirement crop during the season and one more (under irrigated condition) in the post season can be raised.

Wet Zone-II: Areas experience the rainy season of more than 150 days and the average precipitation intensity of more than 10 mm/day are classified in this type. Two successive crops with high water demand can be raised in these zones.

Histogram of the mean monthly rainfall of the six zones is provided in figure 14 to facilitate quick visual comparison of time-distribution of the annual rainfall giving rise to rainy season of so widely different durations over different parts of the country. The histogram for all-India is given to get a feel of average rainfall time-distribution and the rainy season. While most of the annual rainfall is concentrated in four months i.e. June through September over NWI, it is well spread over nine months, April through December over SPI. The remaining four zones in order NCI, WPI, EPI and NEI can be regarded as intermediate cases between these two extremes. For the entire country it is well spread over 7-8 months, April/May through October/November.

6.2.3.2. Summer Monsoon and Northeast Monsoon Rainfall Series Considered

Studies in the past have identified several regional and global circulation features that show significant correlation with the all-India summer monsoon rainfall variations. It is investigated whether the circulation parameters can be used for regional scale prediction of the summer monsoon and all-India and regional scale prediction of the northeast monsoon rainfall. The summer monsoon rainfall series of desired length, depending upon length of the circulation parameter, is adopted from Sontakke and Singh (1996) and the northeast monsoon rainfall series from Singh and Sontakke (1999) who have reconstructed the longest possible instrumental all-India and zonal summer monsoon and northeast monsoon rainfall series using optimum observations.

6.3. Regional/Global Circulation Parameters

To know possible cause(s) of variations in the summer monsoon rainfall over India one traditional approach has been to examine changes in the atmospheric/oceanic phenomena vis-a-vis changes in the monsoon rainfall through correlation or composite analysis. This led

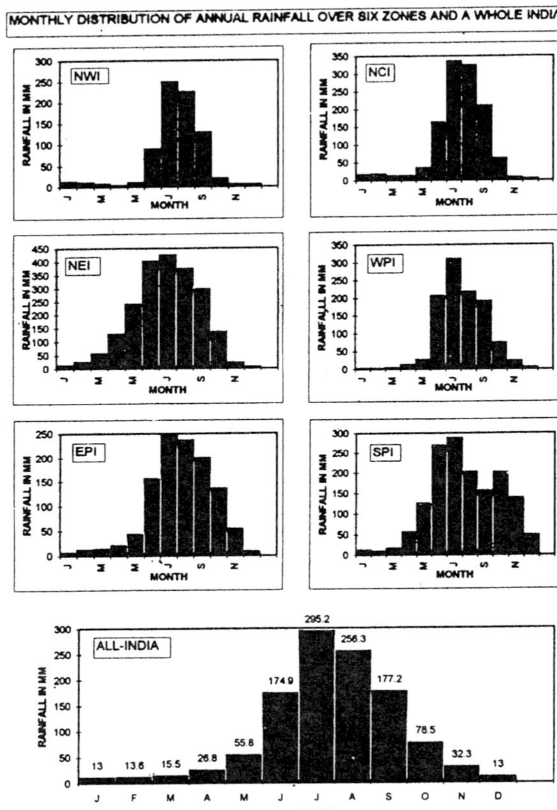

Fig. 14: Histogram of the normal monthly rainfall of the six zones and the whole country.

to the discovery of numerous phenomena which either affect or are affected by the Indian summer monsoon. The following twenty selected parameters are considered for the present investigation (except the individual or group indicated against the parameter they all were earlier attempted by Mooley *et al.* 1985; Parthasarathy *et al.* 1988 & 1990a and others).

1. Sea level pressure (SLP) at Mumbai (1941-1990)
2. Sea level pressure (SLP) at Darwin (1941-1990)
3. Sea level pressure (SLP) at Tahiti (1941-1990)
4. Sea level pressure (SLP) at Cape Town (1951-1986)
5. Sea level pressure (SLP) at Agalega (1941-1990)
6. Sea level pressure (SLP) at Nouvelle (1941-1990)
7. Sea level pressure (SLP) at six selected stations over West Central India (1941-1990)

8. SLP difference between South America and Mumbai (1951-1986)
9. SLP difference between Tahiti and Darwin (1941-1990)
10. SLP difference between Tahiti and Bombay (1941-1990)
11. Northern hemisphere surface air temperature (Verma *et al.*, 1985) (1941-1987)
12. Surface air temperature over India (Parthasarathy *et al.*, 1990b) (1941-1988)
13. Surface air temperature over Pakistan (Singh and Sontakke, 1996b) (1941-1993)
14. Sea surface temperature (SST) at Niño West 160°E–150°W) (1940-1986)
15. SST over the Indonesian region (Nicholls, 1995) (1949-1991)
16. SST at Puerto Chicama (Parthasarathy and Sontakke, 1988) (1941-1989)
17. Frequency of westerly wind days over Britain (Singh, 1995) (1941-1989)
18. 50-mb quasi-biennial oscillation (QBO) at Balboa (Singh, 1995) (1951-1987)
19. 30-mb QBO at Balboa (Mukherjee *et al.*, 1985) (1951-1987)
20. 10-mb QBO at Balboa (Bhalme *et al.*, 1987) (1958-1986)

Admittedly readily available data was also a limiting factor in the choice of these parameters. The parameters from Sl. No. 1 to 10 are used as an index (quantitative measure of the strength) of southern oscillation (SO), from 11 to 13 as a measure of thermal conditions of the northern hemisphere/regional land mass, from 14 to 16 severity of El Niño/La Niña, the 17th the strength of the extra-tropical westerlies and from 18 to 20 the phase of the quasi-biennial oscillation (QBO) of the equatorial lower stratospheric zonal wind. A brief description of the mechanism of different phenomena is given in order. Notwithstanding the El Niño and the southern oscillation are the oceanic (El Niño) and atmospheric (SO) manifestations of the same coupled ocean-atmosphere-land phenomena, popularly known as ENSO, their discriptions are given together.

6.3.1. El Niño and Southern Oscillation (ENSO)

In the last about three decades numerous theories have been proposed to provide plausible explanation of the mechanism of El Niño-Southern Oscillation cycle (Barnett, 1984; Bjerknes, 1969; Meehl, 1994; Tomita and Yasunari, 1996; Webster and Yang, 1992; Yasunari, 1987). The present understanding is that ENSO is a continuously evolving coupled land-atmosphere-ocean phenomenon involving interactions between tropical and extra-tropical atmospheric processes over oceanic and land surfaces of the near-global domain, truly an integral component of the earth system climate dynamics. We briefly describe only two widely accepted theories, the one by K. Wyrtki and another by T. Tomita and T. Yasunari.

Wyrtki Hypothesis: The occasional appearance of warm water off the coast of Peru during the beginning of the year is known as El Niño (Wyrtki, 1975). The 'seesaw' type association between the pressure fluctuations over the Australian-Indonesian region and the southeast Pacific was named Southern Oscillation by its discoverer Walker (Walker, 1924).

Wyrtki (1982) propounded a theory of the ENSO mechanism. His explanation starts from the premise when pressure is high at Easter Island (27°10'S, 109°26'W) and it is low over the Indonesian region. In this situation, the southeast trade winds, equator upwelling and south equatorial current are strong. The result is an accumulation of warm water, a depression of the thermocline and an increase of sea level in the western Pacific, while a lowering of temperature in the eastern equatorial Pacific, especially off the coast of Peru. The region in and around Indonesia experiences low level atmospheric convergence, large-scale convective acti-vities, rising air and heavy rainfall. The large temperature diffe-rence between the east and west enhances the atmospheric circu-lation, particularly the 'Walker Cell'. The strong Walker circulation dominates the atmospheric situation resulting in a weak Hadley circulation. The Hadley cells feed moist air from both hemispheres toward the inter-tropical convergence zone, where air rises, and the moisture is released as rain. At higher levels the air returns poleward to the subtropics, where it sinks and completes the cycle. A weak Hadley circulation also implies a weak vorticity transport from the tropics to the subtropics and causes a weakening of the subtropical high pressure system. Once the subtropical highs, especially the Easter Inland High, become weak, the trade winds relax and an El Niño event will be triggered. A new convective region develop over South America giving massive rains and floods in normally arid countryside. This results in a warm Peru Current, a warm equatorial current, and a weakening of the Walker circulation.

The collapse of the wind field also causes an equatorial Kelvin wave to pass from the western to the eastern Pacific in about 60 days. The energy for the Kelvin wave is provided by the accumulated warm water in the western Pacific and the collapse of the winds only triggers it. When the Kelvin wave impinges on the coast of South America, downwelling occurs and warm waters accumulates off the coast. The south equatorial currents decrease in strength, so cool advection in the Peru current decreases. Equatorial upwelling decreases both because of the Kelvin wave and the weaker winds, and consequently equatorial waters warm rapidly. The temperature difference between east and west is small and as a result the Walker circulation is weak. The Walker circulation together with warmer equatorial waters leads to a stronger Hadley circulation and a stronger vorticity transport to the subtropics, which in turn intensifies the subtropical high and closes the cycle. The actual El Niño event lasts only 14 to 16 months, but the cycle may be as fast as three years and as long as a decade (Wyrtki, 1982).

Later Wyrtki (1985) suggested that during prolonged atmospheric circulation with normal strength over the tropics the trade winds push warm water toward west and cause it to accumulate in the western Pacific. This process continues several years. Atmospheric fluctuations sometimes lead to a relaxation of the trade wind field which allows triggering of a Kelvin surge, viz. a massive eastward displacement of the accumulated warm water along the equator. The displacement of warm water is accompanied by shift of area of strong convection and rainfall from Indonesian region to the central and eastern Pacific. The warm water is deflected by the coast of America and is lost to the higher latitudes of both the hemispheres. At the end the tropical ocean is depleted of heat, which is restored by the slow accumulation of warm water in the western Pacific by the normal trade winds. The time scale of the Southern Oscillation is given by the time required for the accumulation of warm water in the western Pacific.

An El Niño-Southern Oscillation cycle represents a heat relaxation of the ocean-atmosphere system, in which heat stored in the tropical ocean is discharged toward higher latitudes. The cycle seems to be a manifestation of complex interaction between random atmospheric processes and deterministic oceanic processes.

TOMITA AND YASUNARI HYPOTHESIS: From global analysis of sea level pressure (SLP), sea level wind, sea surface temperature (SST) and outgoing longwave radiation (OLR) Tomita and Yasunari (1996) have demonstrated that tropical-extratropical interactions through the northern winter northeast monsoon circulation play an important role in the ENSO-monsoon mechanism. In situations when the Indian southwest monsoon is weaker, the convective activity around the Indonesian maritime continent and the Walker Circulation are also weak. There is an occurrence of westerly wind anomalies over the tropical central western Pacific and excitation of El Niño like situation. The positive SST anomalies in the central and eastern tropical Pacific develop during the period from the northern summer to winter. The positive SST anomalies excite the Rossby wave which forms the Pacific-North American (PNA) pattern in the northern winter (Horel and Wallace, 1981). The modulated SLP anomalies of this PNA pattern form the large positive anomaly centered over the central north Pacific. The atmospheric circulation associated with this anomaly centre appears to be coupled with the weakening of the northeast winter monsoon (NEWM) which causes positive SST anomaly around the South China Sea due to increased insolation. After the NEWM ends, the seasonal migration of convective activity around the maritime continents is strengthened by the positive SST anomalies around the South China Sea. Stronger convection may contribute to stronger Indian southwest monsoon activity. The La Niña like phenomenon occurs due to the intensification of the Walker Circulation. The La Niña phase persists to the next northern spring, with a similar seasonal processes but with the reversed signs of the anomaly patterns and eventually the cycle of the biennial oscillation of the ENSO-monsoon system is completed. Further evidence has also been provided that the NEWM is closely related to the Eurasian pattern and western Pacific patterns of the northern hemisphere 500 hPa height anomalies. The weakening/strengthening of the Indian summer monsoon is associated with the Eurasian pattern, Eurasian snow cover and the circulation features of the northern hemisphere.

A conceptual visualization of the El Niño cycle in a QBO mode and the associated changes in the Indian summer monsoon circulation features is summarized in Table 2.

Table 2: El Niño and the Indian summer monsoon relationship in a QBO mode

1.	Warmer N.H. surface air temp. ⇓	1.	Cooler N.H. surface air temp. ⇓
2.	Higher N.H. 500hPa contour height ⇓	2.	Lower N.H. 500hPa contour height ⇓
3.	Stronger upper tropospheric westerlies with relatively long, small amplitude waves restricted to mid-latitudes ⇓	3.	Weak upper tropospheric westerlies with relatively short, large amplitude waves tend to shift southwards into subtropical, or even tropical, latitudes ⇓
4.	Well developed and strong Tibetian Anticyclone ⇓	4.	Temporary partial distruction of the Tibetian High or the Tibetian High is poorly developed over East/Southeast of its normal position ⇓
5.	Strong tropical easterly jet (TEJ) ⇓	5.	Weak tropical easterly jet (TEJ) ⇓
6.	Strong cross equatorial flow along Somali coast ⇓	6.	Weak cross equatorial flow along Somali coast ⇓
7.	Strong low level westerly/southwesterly winds over Arabian Sea peninsular India ⇓	7.	Weak low level westerly/southwesterly winds over peninsular India ⇓
8.	Well developed and intense monsoon trough over Indo-Gangetic plains of India ⇓	8.	Weak and poorly developed monsoon trough over Indo-Gangetic plains ⇓
9.	Intense and well developed seasonal 'heat low' over northwest India-Pakistan-Afghanistan-Iranian sector ⇓	9.	Weak and poorly developed seasonal 'heat low' over northwest India-Pakistan-Afghanistan-Iranian sector ⇓
10.	Strong convective activity over the Indian ocean ⇓	10.	Less convective activity over Indian ocean ⇓

11.	Strong Walker circulation with ascending branch over India and neighbourhood and descending branch over Pacific region ⇓	11.	Weak Walker circulation with descending branch over India and neighbourhood and ascending branch over Pacific region ⇓
12.	Heavy rainfall activity over the Indian region ⇓	12.	Lesser rainfall activity over Indian region ⇓
13.	Stronger trade winds over the tropical Indian and Pacific oceans ⇓	13.	Weaker trade winds or outburst of westerly winds over tropical central-western Pacific ⇓
14.	Low SST, high pressure and weak convective activity over the Southeast Pacific region ⇓	14.	High SST, low pressure and strong convective activity over the Southern Pacific region ⇓
15.	Intense Hawaiin High and North Pacific pattern between Hawaiin High and Aleutian Low ⇓	15.	Weak Hawaiin High and North Pacific pattern between Hawaiin High and Aleutian Low ⇓
16.	Strong cold air advection over the Eurasian landmass ⇓	16.	Weak cold air convection over the Eurasian landmass ⇓
17.	Decrease in the surface air temp. of the N.H. ⇓	17.	Increase in the surface air temp. of the N.H. ⇓

6.3.2. Northern Hemisphere Temperature

State of thermal condition of the northern hemisphere play an important role in the occurrence of the Asian southwest monsoon and its variability. Winstanley (1973) hypothesized that during warmer northern hemisphere circumpolar westerlies of the upper troposphere are contracted polewards and the monsoon currents penetrate more than usual northward giving a higher rainfall over northwest India. During cooler northern hemisphere a weak westerlies with relatively short, large amplitude waves tend to shift southwards into subtropical, and even tropical, latitudes, the tropical meridional circulation contracts and the monsoon rains do not extend as far north. Verma *et al*, (1985) have reported for the first time a significant positive correlation between the Jan-Feb surface air temperature of the northern hemisphere (NH) (Jones *et al*, 1986) and the all-India SMR in recent period. In a detailed examination Angell and Korshover (1978) have shown an inverse relation between the areal extent of the north circumpolar vortex (area north of the 300-mb contour in the main belt of westerlies on the mean monthly polar stereographic maps) and mid-latitude troposphere temperature.

6.3.3. Temperature over India and Pakistan

The 'low' over the hottest regions of the southern parts of Arabia, Pakistan, Baluchistan and adjoining northwest India play a crucial role in the occurrence and variation of the summer monsoon circulation. The pressure gradient from the south Indian Ocean high to this low produces a strong lower tropospheric cross-equatorial flow and the monsoon winds. Higher the temperature, lower is the pressure. This simple logic tempted the researchers to examine the association between the temperature over and/or adjoining the 'low' pressure area and the large scale features of the summer monsoon rainfall (recently Mooley and Paolino, 1988; and Parthasarathy *et al*, 1990a and others).

6.3.4. Frequency of Westerly Wind Days over Britain

Highly mobile westerly weather systems giving mild, wet winters over England are linked to a strong zonal westerly circulation in the mid-tropospheric circumpolar westerlies (Lamb, 1972). Winstanley (1973) demostrated a close relationship between the strength of the zonal circulation and rainfall over England and an equally close relationship with the summer monsoon rainfall over the Indian subcontinent. During strong zonal circumpolar westerlies highly mobile depressions and ridges travel eastwards across the Atlantic to give changeable weather and high rainfall over England. Meridional circulation in the tropics is strong favouring more northerly penetration of the summer monsoon currents and greater rainfall. During weaker zonal westerlies, frequency of westerly winds and rainfall over Britain decrease. Westerly trough extend into subtropical and even tropical latitudes and rainfall over the Mediterranean and the Middle East is higher. The tropical meridional circulation contracts, and monsoon rains do not extend as far north. In the quantitative analysis, however, Singh (1995) has found only an isolated significant correlation of –0.44 between the May westerly wind days and the square-root of the summer monsoon dry area (SMDA) series (1931-1960).

6.3.5. Quasi-Biennial Oscillation of the Equatorial Lower Stratosphere

In the equatorial stratosphere from 60 to 10 mb alternate spells of easterly and westerly winds occur with a period of about two years (occasionally of longer duration) known as quasi-biennial oscillation (QBO). The wave-mean flow interaction is attributed as the cause of the stratospheric QBO (Nicholls, 1980). Some of the studies which investigated the association between the QBO and the Indian summer monsoon are Mukherjee *et al* (1985); Bhalme *et al* (1987); Singh (1995) and others.

The correlation coefficient between 3-month running mean from the preceding one year through concurrent to the following one year of each of the twenty parameters and the all-India and the zonal summer monsoon as well as northeast monsoon rainfall is calculated using data of the indicated period. For the sake of brevity only the result for the all-India SMR is displayed in Figure 15, and the significant values (at least at 5%. level) for the all-India summer monsoon as well as northeast monsoon rainfall are given in Tables 3 and 4, respectively. However, a brief description is provided in the following section for the all-India and the zonal rainfall of both the monsoons.

Table 3: Summary of the correlation analysis of the 19 circulation parameters (Col. 2) with the summer monsoon rainfall of the whole country. Col. 3 lists the period(s) in the preceding one year (-1 yr) through the concurrent year (0 yr) to the following one year (+1 yr), during which the 3-month running means of the particular parameter have shown significant correlation; the Col. 4 gives the range of the correlation coefficient (CC) values; and Col. 5 the level of the statistical significance.

Sr. No.	Circulation parameter	Period(s) that showed significant correlation	Correlation Coefficient (CC)	Significance level
1.	SLP at Mumbai	-1 yr Feb-Jun; 0 yr Mar-Nov	0.33 to 0.38; -0.32 to -0.72	5% to 1%; 5% to 0.1%
2.	SLP at Darwin	-1 yr Apr-Aug; -1 yr Nov-0 yr Feb; 0 yr Mar-+1 yr Apr	0.33; 0.33; -0.34 to -0.59	5%; 5%; 5% to 0.1%
3.	SLP at Tahiti	0 yr May-+1 yr May	0.33 to 0.48	5% to 1%
4.	SLP at Capetown	-1 yr Sept-Nov; 0 yr Aug-Oct; 0 yr Dec-+1 yr Feb	-0.40; 0.38; -0.42	5%; 5%; 5%
5.	SLP at Agalega	-1 yr Jan-Apr; 0 yr Jun-Oct; 0 yr Dec-Mar	0.30 to 0.34; -0.31 to -0.40; -0.33 to -0.38	5%; 5% to 1%; 5%
6.	SLP at Nouvelle	0 yr Jun-Oct; +1 yr Mar-May	-0.28 to -0.38; 0.40	5% to 1%; 1%
7.	SLP at WC-India	-1 yr Feb-Jun; 0 yr Mar-Nov	0.34 to 0.41; -0.42 to -0.69	5% to 1%; 0.1%
8.	SLP S. America-Mumbai	0 yr Apr-Nov; +1 yr Jan-Jun	0.37 to 0.63; 0.39 to 0.42	5% to 1%; 5%
9.	SLP Tahiti-Darwin	-1 yr Mar-Aug; 0 yr Apr-+1 yr May	-0.31 to -0.43; 0.38 to 0.61	5% to 1%; 1% to 0.1%
10.	SLP Tahiti-Mumbai	-1 yr Feb-Jun; 0 yr Apr-+1 yr May	-0.32 to -0.44; 0.32 to 0.69	5% to 1%; 5% to 0.1%
11.	N.H. Temp.	-1 yr May-Oct; -1 yr Nov-0 yr Mar; 0 yr Dec-+1 yr Sept	0.30 to 0.45; 0.31 to 0.35; -0.29 to -0.49	5% to 1%; 5%; 5% to 0.1%
12.	Indian Temp.	-1 yr May-Sept; 0 yr May-+1 yr Jul	0.35 to 0.44; -0.36 to -0.68	5% to 1%; 5% to 0.1%
13.	Pakistan Temp.	-1 yr Jun-Oct; 0 yr Feb-Jul	0.33 to 0.46; 0.32 to 0.39	5% to 0.1%; 5% to 1%
14.	SST Nino West	0 yr Dec-+1 yr Apr	0.31 to 0.37	5%
15.	SST Indonesia	0 yr Jan-Dec.	0.31 to 0.46	5% to 1%
16.	SST Puerto Chicama	-1 yr Jan-Apr; 0 yr Jun-+1 yr Feb	0.29 to 0.31; -0.33 to -0.53	5%; 5% to 0.1%
17.	50 mb - QBO	-1 yr May-Nov; 0 yr Jul-+1 yr Apr; +1 yr Sept-Dec	-0.30 to -0.50; 0.33 to 0.40; -0.42 to -0.47	5% to 1%; 5%; 1%
18.	30 mb - QBO	-1 yr Jan-Jul; 0 yr Mar-Oct; +1 yr Jun-Dec	-0.35 to -0.43; 0.33 to 0.42; -0.38 to -0.43	5% to 1%; 5% to 1%; 5% to 1%
19.	10 mb - QBO	-1 yr Jul-0 yr Apr; 0 yr Sept-+1 yr Apr; +1 yr Sept-Dec	0.41 to 0.59; -0.48 to -0.55; 0.45 to 0.48	5% to 0.1%; 5% to 1%; 5%

SLP:Sea Level ro; SST:

Table 4: Same as in Table 2, but for the northeast monsoon rainfall over whole India

Sr. No.	Circulation parameter	Period(s) that showed significant correlation	Correlation Coefficient (CC)	Significance level
1.	SLP at Mumbai	0 yr Sept-+1 yr Jan	-0.29 to -0.39	5% to 1%
2.	SLP at Tahiti	0 yr Apr-Jul	-0.29 to -0.40	5% to 1%
3.	SLP at Nouvelle	-1 yr Dec-+0 yr Feb; +1 yr Jun-Sept	-0.33; -0.30 to -0.41	5%; 5% to 1%
4.	SLP at WC-India	0 yr Sept-Nov	-0.30	5%
5.	SLP S. America-Mumbai	0 yr Mar-Jun; 0 yr Aug-+1 yr Jan	0.37 to 0.40; 0.38 to 0.50	5%; 5% to 1%
6.	SST Nino West	0 yr May-Jul	-0.38	5%
7.	10 mb - QBO	-1 yr Nov-0 yr Jun	0.37 to 0.56	5% to 1%

Although the objective is to explore the predictive potential of summer monsoon and northeast monsoon rainfall over the traditional rice growing areas, the NWI and the WPI are included to see how the results compare with the dryland areas, and the all-India is included to facilitate comparison with the earlier published works.

6.4. Results of the Correlation Analysis

6.4.1. Summer Monsoon Rainfall

The correlation of 3-month running mean from preceding one year through concurrent to the following one year of the twenty selected circulation parameters with the all-India series is displayed in Figure 15, and its 5% significant value documented in Table 3. As stated previously, the parameters are chosen based on reports of the earlier studies that they show significant correlation with the all-India SMR. Some variation in the present CC from its value reported earlier is due to different set of rain gauges and methods used in preparing the all-India series, and also the different data period and its length considered in the analysis. Predictive model can be developed by screening these physically-based correlated parameters, but the value of prediction of average rainfall over such a large region is limited.

North West India

Different indices of SO, SATs, SSTs and equatorial lower stratospheric zonal wind at 50 mb, 30-mb and 10-mb levels observed over Balboa have shown significant correlation. This is understandable. The SMR series of North West India is most highly correlated with the all-India SMR series, and hence these circulation parameters. Development of predictive model for the NWI can be a parallel program alongwith the whole country.

North Central India

The SLP at Mumbai, Darwin, Capetown, Nouvelle, west coast of India, South America minus Mumbai and Tahiti minus Mumbai, SATs, SSTs over the Indonesian region and Puerto Chicama, and the Balboa zonal wind at 50-mb, 30-mb and 10-mb showed significant correlation. An adequate number of parameters preceding the summer monsoon season provides an indication that development of predictive model for the area can be attempted.

North East India

The SLP at Mumbai, Capetown, Agalega, Nouvelle, west coast of India, South America minus Mumbai, Tahiti minus Darwin and Tahiti minus Mumbai, SATs, SSTs and the Balboa lower stratospheric zonal wind showed significant correlation. Limited parameters such as SLP difference between South America and Mumbai, SAT of Pakistan and that of the northern hemisphere and the Balboa 10-mb zonal wind preceding the summer monsoon season would require skillful choice of the technique for developing the predictive model.

West Peninsular India

Different pressure-based SO indices, SATs, SST of the Niño West region and that of Puerto Chicama and the Balboa 10-mb zonal wind have shown significant correlation. Second to NWI, the WPI series shows highest correlation with the all-India series. Predictive model for the area can be developed almost with the same set of independent parameters that are in NWI and all-India models.

East Peninsular India

The SO indices (except SLP at Darwin and Nouvelle), SATs, SSTs over Niño West and that over Indonesian region, and the Balboa zonal wind at all the three levels showed significant correlation. Alongwith these the Balboa 10 mb zonal wind from –1 yr July-to-0 yr June, the concurrent year Mar-Apr-May SLP of the west coast of India and the preceding year Jan-Feb-Mar SAT of Pakistan can provide sufficient number of independent parameters for developing predictive model of the area.

South Peninsular India

Different SO indices, SATs, SST of the Indonesian region and that of the Puerto Chicama and the Balboa 50 mb and 10 mb zonal winds, all generally of the concurrent period, showed significant correlation. SLP at Capetown, SLP difference between Tahiti and Mumbai, northern hemisphere temperature, Pakistan temperature and the Balboa 10 mb zonal wind of the concurrent/preceding year can provide important independent parameters for the predictive model of the area.

6.4.2. Northeast Monsoon Rainfall

Only seven parameters that is SLP at Mumbai, Tahiti, Nouvelle and west coast of India, SLP difference between South America and Mumbai and Tahiti and Mumbai, SST over Niño West area and the Balboa 10 mb zonal wind showed significant correlation (Table 4). Except the Tahiti SLP and the Niño West SST the different parameters have shown the correlation of the same sign as with the all-India SMR. The predictive model for the all-India northeast monsoon rainfall is not expected to be as reliable as for the all-India SMR because of lesser number of parameters and their weaker correlations.

North West India

The SLP at Mumbai, Darwin, Capetown, Nouvelle, west coast of India, the SLP difference between South America and Mumbai, Pakistan temperature, SST over Niño West area and that over Puérto Chicama and the Balboa 10 mb zonal wind showed significant correlation. A good number of parameters that have shown significant correlation with the summer monsoon also showed significant correlation with the NEMR of the area. But in case of northeast monsoon the period(s) of significant correlation occurred much before the start of the season. Thus development of model giving much earlier prediction can be attempted.

North Central India

The SLP at Mumbai, Darwin, Capetown, Nouvelle, west coast of India, and SLP difference between South America and Mumbai and the Balboa 10 mb zonal wind showed significant correlation. Interesting to note is that except SLP at Nouvelle the period of significant correlation for different parameters occurred much earlier to the season. But reliability of predictive model seems to be limited.

North East India

Contrary to popular belief much larger number of parameters have shown significant correlation with the northeast monsoon rainfall over the area. The parameters are SLP at Mumbai, Capetown, Agalega and west coast of India, SLP difference between South America and Mumbai, Tahiti and Darwin, and Tahiti and Mumbai, SAT over India and that over Pakistan, SST of the Indonesian region and the Balboa 50 mb zonal wind. In respect of lead time the occurrence period of the significantly correlated is much earlier compared to the SMR of the area. However, sound predictive model for the area does not seem possible.

West Peninsular India

The parameters that showed significant correlation in the concurrent and preceding periods are SLP at Mumbai, Capetown and

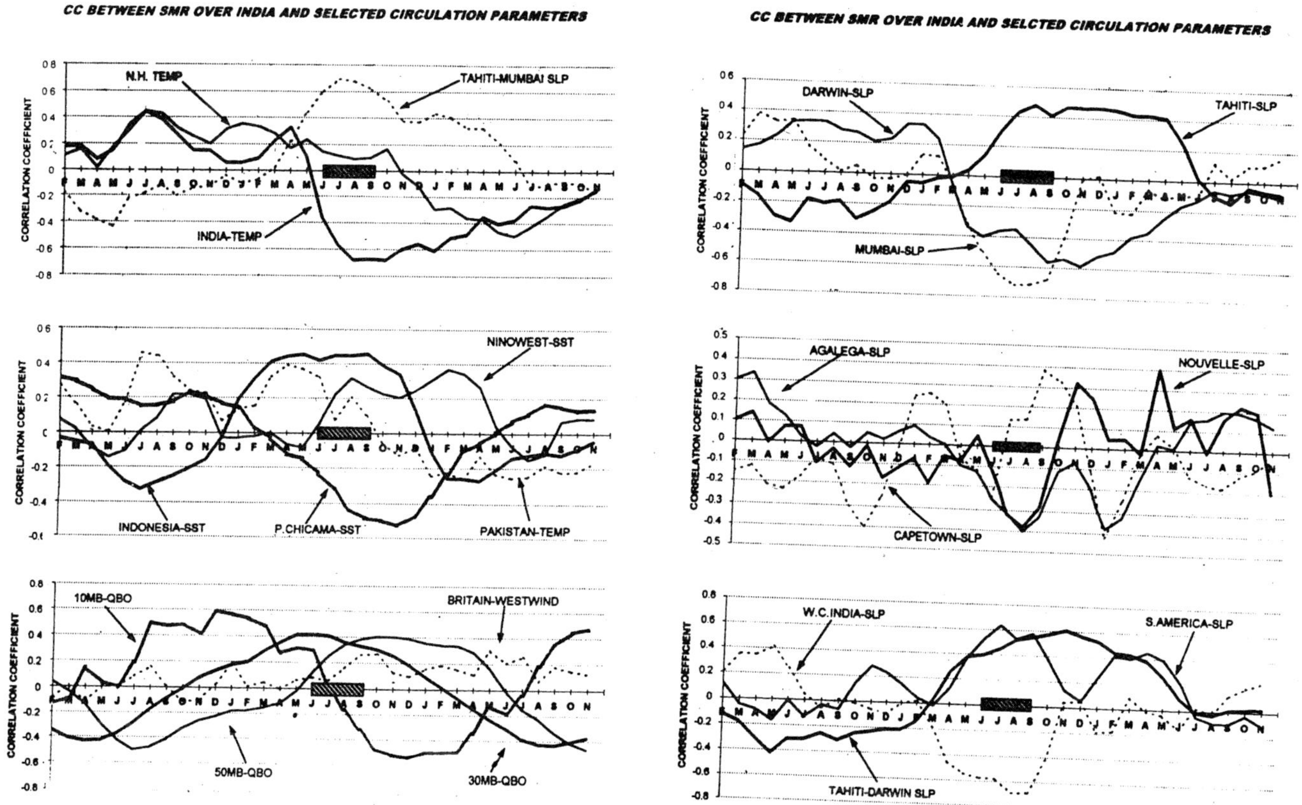

Fig. 15: The correlation coefficient (CC) between 3-month running mean from the preceding one year through contemporary to the following year of the indicated circulation parameter and the all-India SMR (data used in the calculation is given in the text).

west coast of India, SLP difference between South America and Mumbai and 50 mb zonal wind over Balboa. Development of predictive model does not seem an easy task.

East Peninsular India

The parameters of the preceding and the concurrent period that showed significant correlation are SLP at Mumbai and that at Tahiti, SLP difference between South America and Mumbai, SAT of India and SST of the Niño West area. Reliable predictive model is not expected.

South Peninsular India

The SLP at Mumbai, Darwin, Tahiti, Agalega and Nouvelle, SLP difference between Tahiti and Darwin and between Tahiti and Mumbai and SAT over India, over Pakistan and over the northern hemisphere showed significant correlation. Some of the parameters such as SLP at Tahiti, SLP difference between Tahiti and Darwin and between Tahiti and Mumbai and the SAT have shown correlation of opposite sign to that shown with the summer monsoon rainfall of the area. To a large extent this is consistent with the normal expectation as there is a reversal in the meanflow of the lower tropospheric wind from southwest to northeast. But pressure, wind and thermal conditions have to be conducive for low level moisture convergence, convective activities, enhanced vertical velocities and rainfall over the Indian region during October through December period. Then how the reversal in the direction of the certain processes keep helping the rainfall activities of the northeast monsoon in the face of the other parameters which show the correlation of the same sign as with the summer monsoon rainfall? Any attempt to get a physical picture of this situation will raise a doubt whether the correlation is realistic or merely a chance. For moderate correlation of around 0.4 change in the sign of correlation with the change in the data period is not unexpected. No case of significant correlation is found with the frequency of westerly wind days over Britain.

The correlation of the 200-mb May meridional wind over India (Joseph, 1978) with the SMR series (1964-1991) of the six zones and all-India in order is –0.57, –0.38, –0.03, –0.52, –0.46, –0.35 and –0.65, and that of April 500-mb ridge along 75°E (Banerjee *et al*, 1978) 0.50, 0.41, 0.18, 0.50, 0.37, 0.20 and 0.61 based on data of 1939-1994. But these two important regional circulation parameters cannot be utilized for northeast monsoon rainfall studies.

6.5. A Time Series Modelling and Extrapolation Approach to Rainfall Prediction

From the correlation study with the circulation parameters it is inferred that in most of the cases attempt to develop reliable regression model for the regional rainfall prediction might be a futile exercise. Even if it succeeded in some case(s), timely non-availability of predictor parameters will always be a handicap to run such models for real time operational forecasting. An alternate scheme, time series modelling and extrapolation, is suggested for operational forecasting after having examined the fluctuation characteristics of the longest instrumental all-India and zonal summer monsoon (Sontakke and Singh, 1996) and northeast monsoon rainfall series (Singh and Sontakke, 1999) which were reconstructed using optimum observations identified from the application of an objective selection technique (Singh, 1994). The technique also known as one-step-ahead-forecasting (OSAF) scheme is described using northeast monsoon rainfall series (1813-1997) of the South Peninsular India as an example. Earlier such approaches to long range weather or short term climate prediction were attempted by Schickedanz and Bowen (1977), Kane (1988), Kane and Teixeiro (1990), Kane and Trivedi (1988) and many others. The main attraction of this sort of approaches is that they do not require any exogeneous information. Abstraction of significant implicit periods is not a necessary requirement, though harmonic decomposition is the first step of the scheme.

Suppose the rainfall series is $[z(t)_n]$: z(1), z(2), z(3),......z(N), the following calculations are carried out in sequential order to develop the final prediction model.

i. Consider a large segment of the actual rainfall time series $[z(t)_{N-m}]$: z(1), z(2), z(3),......z(N-m), where m (say about 30 to 35) denotes the number of latest data points to be used for comparison between predicted and actual values.

ii. Fill the (N–m+1)th term by the mean $\bar{z}(t)_{N-m}$.

iii. Decompose the entire sequence $z(t)_{N-m+1}$ (actual plus the (N–m+1)th term filled with mean) into all possible harmonics.

iv. Arrange the harmonics in order from the highest to the lowest variance contribution.

v. Regenerate the entire sequence by combining the first few harmonics such that the correlation between generated and the actual sequence is ~ 0.70 (adjusted after few trials by changing the number of harmonics).

vi. Pick up the generated amount at the (N–m+1)th term.

vii. Enter the second round of calculation by considering the segment of increased length by one term (N–m+1) from the original sequence, i.e. $[z(t)_{N-m+1}]$: z(1), z(2), z(3),......z(N–m+1).

viii. Fill the (N–m+2)th term by the mean $\bar{z}(t)_{N-m+1}$, decompose the entire sequence into all possible harmonics, arrange the harmonics in decreasing order of the variance contribution, regenerate the sequence by combining the first few harmonics which shows correlation of ~ 0.7 with the actual sequence, and pick up the regenerated amount at the (N–m+2)th term.

ix. Continue the process considering one additional term from the original sequence.

x. Terminate the process at the stage the entire original sequence $[z(t)_n]$ is involved in the calculation and the rainfall amount for the (N+1)th term generated.

xi. Estimate the least-squares simple linear regression between the generated rainfall amount at the terms $z(N-m+1)$, $z(N-m+2)$......$z(N)$ and their corresponding actual values.

xii. Substitute the generated amount at the (N+1)th time step in the regression and get the prediction for the time step.

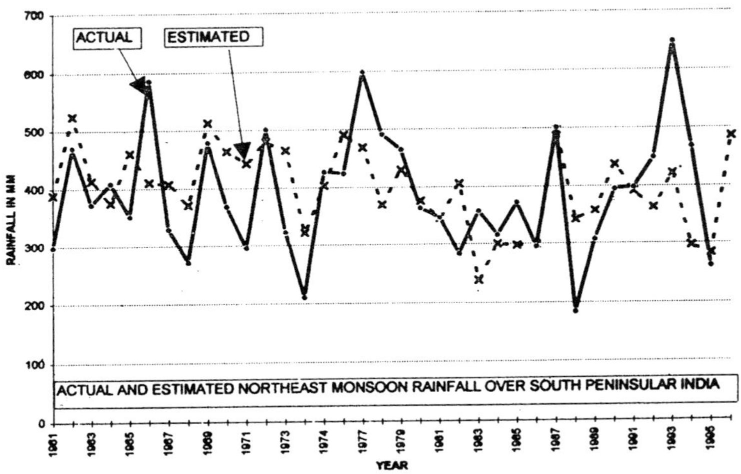

Fig. 16: Actual and estimated northeast monsoon rainfall of the South Peninsular India for the period 1961-1997. Predicted amount for the year 1998 is also shown.

The following calculations in order are carried out to develop the predictive model for the NEMR over the South Peninsular India (SPI) and make the prediction for the year 1998. The available long period (1813-1997) reliable reconstructed NEMR series for the region is utilized for the purpose.

i. Consider the series of the period 1813-1960, substitute the mean of the series $\bar{z}_{1813-1960}$ for the year 1961, decompose the series 1813-1961 into all possible harmonics, arrange the harmonics in the order of variance contribution from highest to lowest, regenerate the series by combining the first few harmonics showing correlation of~ 0.70 with the actual series (1813-1960), and get the generated rainfall amount for the year 1961.

ii. In the second round consider the series of the period 1813-1961 and repeat the different calculations of step (i) including generation of rainfall amount for the year 1962.

iii. Continue the process considering one additional term in each new round of calculations.

iv. Terminate the process when the entire actual rainfall sequence (1813-1997) is considered and the rainfall amount for the year 1998 is generated from the combination of selected few harmonics.

v. Estimate the linear regression between the generated values (X_G) of the period 1961-1997 and their corresponding actual values (X_A). The regression for the NEMR over SPI is as follows ($R = 0.76$)

$$X_A = -234.1 + 1.6\,X_G$$

Substitute the generated value for the year 1998 in the above equation and get the prediction for the year. According to the present approach the NEMR over SPI during 1998 is likely to be about 4.4% below normal, 369 mm against the normal rainfall of 386 mm. The actual and estimated rainfall amounts for the period 1961-1997 and the predicted amount for the year 1998 are shown in Figure 16.

6.6. Concluding Remarks

Contrary to the popular belief a large number of circulation parameters showed significant correlation with regional summer monsoon rainfall and with regional and all-India northeast monsoon rainfall (NEMR), the choice of the parameters was solely based on their reported significant correlation with the all-India summer monsoon rainfall. Though the correlation is generally weak, the present preliminary survey is expected to open new avenue to try for other circulation parameters which can be helpful to better understand the mechanism of the rainfall processes over the particular region, and in turn reliable physical-based predictive rainfall model.

We believe that reliability of the present time series modelling and extrapolation approach to rainfall prediction would depend upon the degree of regularity in the rainfall variations, greater the regularity, that is if fewer wave forms explain greater portion of the variance, better the prediction.

Almost every terrestrial and extra-terrestrial parameter displays certain amount of regularity in its variation. But at what time scale of averaging or totalling the regularity emerges the pest needs to be determined for every parameter. In case of rainfall, even the time scale can show large spatial variability. For example, at some places 5-day total can show best regularity, but it may be 10-day for some other place, 30-day for a third place etc. There does not seem to exist any objective criterion to obtain the time scale at which the best regularity appears in the rainfall of a given station. So presently we can try conventional 5-day, 10-day, 15-day, 30-day or seasonal totals to get the rainfall time series showing the best possible regularity.

Acknowledgements

The authors express their gratitude and indebtedness to Dr. G.B. Pant, Director of the Indian Institute of Tropical Meteorology, Pune for the necessary facilities to carry out the work. Cooperation by the India Meteorological Department, Pune in acquiring the rainfall data used in this study is also thankfully acknowledged.

6.7. References

ANANTHAKRISHNAN, R. and PATHAN, J.M. 1991. A climatological singularity around mid-August in the summer rainfall of India. *Curr. Sci.* **60:** 439-445.

ANGELL, J.K. and KORSHOVER, J. 1978. The expanded north circumpolar vortex of 1976 and winter of 1976-77 and attendant vortex displacement, *Mon. Weath. Rev.* **106:** 137-142.

BANERJEE, A.K., SEN, P.N. and RAMAN, C.R.V. 1978. On foreshadowing southwest monsoon rainfall over India with mid-tropospheric circulation anomaly of April, *Indian J. Met. Hydrol. Geophys.* **29:** 425-431.

BARNETT, T.P. 1984. Interaction of the monsoon and trade wind system at inter-annual time scale, Part III: A partial anatomy of the Southern Oscillation, *Mon. Wealth. Rev.* **112:** 2388-2400.

BHALME, H.N., RAHALKAR, S.S. and SIKDAR, A.B. 1987. Tropical quasi-biennial oscillation of the 10-mb wind and Indian monsoon rainfall-implications for forecasting, *J. Climatol.* **7:** 345-353.

BJERKNES, J. 1969. Atmospheric teleconnections from the equatorial Pacific, *Mon. Weath. Rev.* **97:** 163-172.

BLANFORD, H.F. 1884. On the connexion of the Himalaya snowfall with dry winds and seasons of drought in India, *Proc. Roy. Soc.* London, **37**: 3-22.

BLANFORD, H.F. 1886. Rainfall of India, *India Meteorological Department Memoirs*, **81**, vol. III, pp. 658.

ELIOT, J. 1902. Monthly and annual rainfall of 457 stations in India to the end of 1900. *Ind. Met. Memoirs*. Vol. XIV. pp. 709.

Gadgil, S., Yadumani and Joshi, N.V. 1993. Coherent rainfall zones of the Indian region. *Intl. J. Climatology*. **13**: 547-566.

GREGORY, S. 1989. Macro-regional definition and characteristics of Indian summer monsoon rainfall, 1871-1985: *Int. J. Climatol.* **9:** 465-483.

HORREL, J.D. and WALLACE, J.M. 1981. Planetary-scale atmospheric phenomena associated with the Southern Oscillation, *Mon. Weath. Rev.* **109**: 813-829.

INDIAN FAMINE COMMISSION. 1880. *Famine Relief Part I: Measures of protection and prevention*. Blue Book, London C-1591 & C-2735.

INDIAN FAMINE COMMISSION. 1898. Report of Government of India. Central Printing Office. Calcutta. pp. 371.

INDIAN FAMINE COMMISSION. 1901. Report of Government of India. Central Printing Office. Calcutta. pp. 207.

JONES, P.D., RAPER, S.C.B., DIAZ, H.F., KELLY, P.M. and WIGLEY, T.M.L. 1986. Northern hemisphere surface air temperature variations: 1851-1984; *J. Climat. Appl. Meteorol.* **25:** 161-179.

JOSEPH, P.V. 1978. Sub-tropical westerlies in relation to large-scale failure of Indian monsoon, *Indian J. Met. Geophys.* **29**: 412-418.

KANE, R.P. 1988. Spectral characteristics of the series of annual rainfall in England and Wales, *Climatic Change.* **12:** 77-92.

KANE, R.P. and Teixeira, N.R. 1990. Power spectrum analysis of the time-series of annual mean surface air temperatures, *Climatic Change.* **17:** 121-130.

KANE, R.P. and Trivedi, N.B. 1988. Spectral characteristics of the annual rainfall series for northeast Brazil, *Climatic Change.* **13:** 317-336.

KATZ, R.W. and PERLANGE, M.B. 1995. Generalization of chain-dependent processes: Application to hourly precipitation. Water Resources Research. **31**: 1331-1341.

LAMB, H.H. 1972. *Climate: present, past and future.* Methuen, London pp. 613.

LOVEDAY, A. 1914. *The history and economics of Indian famines.* G. Bell and Sons Ltd. pp. 159.

MADDEN, R.A. 1981. Estimates of the natural variability of time-averaged sea-level pressure. *Mon. Wea. Rev.* **104**: 942-952.

MEEHL, G. 1994. Coupled land-ocean-atmosphere processes and south Asian monsoon variability, *Science.* **266**: 263-267.

MOOLEY, DIWAKAR A. 1973. Gamma distribution probability model for asian summer monsoon monthly rainfall. *Mon. Weath. Rev.* **101**: 160-176.

MOOLEY, D.A. and PAOLINO Jr., D.A. 1988. A predictive monsoon signal in the surface level thermal field over India, *Mon. Weath. Rev.* **116**: 256-264.

MOOLEY, D.A., PARTHASARATHY, B. and SONTAKKE, N.A. 1985. Relationship between all-India summer monsoon rainfall and southern oscillation/eastern equatorial Pacific sea surface temperature, *Proc. Indian Acad. Sci. (Earth Planet Sci.)* **94**: 199-210.

MUKHERJEE, B.K., INDIRA, K., REDDY, R.S. and RAMANA MURTHY, Bh. V. 1985. Quasibiennial oscillation in stratospheric zonal wind and Indian monsoon, *Mon. Weath. Rev.* **113**: 1421-1424.

MUTHIAH, S. and RAMACHANDRAN, R. (eds.) 1990. *An Atlas of India*, Oxford University Press, pp. 185.

NICHOLLS, N. 1980. Long range weather forecasting: value, status, and prospects, *Rev. Geophys. Space Phys.* **18:** 771-788.

NICHOLLS, N. 1995. All-India summer monsoon rainfall and sea surface temperatures around northern Australia and Indonesia, *J. Climat.* **8:** 1463-1467.

PARTHASARATHY, B. and SONTAKKE, N.A. 1988. El Niño/SST of Puerto Chicama and Indian summer monsoon rainfall: statistical relationship, *Geofisica International.* **27**: 37-59.

PARTHASARATHY, B. and YANG, S. 1995. Relationships between regional Indian summer monsoon rainfall and Eurasian snow cover, *Adv. Atmos. Sci.* **12**: 143-150.

PARTHASARATHY, B., DIAZ, H.F. and EISCHEID, J.K. 1988. Prediction of all-India summer monsoon rainfall with regional and large-scale parameters, *J. Geophys. Res.* **93**: 5341-5350.

PARTHASARATHY, B., RUPA KUMAR, K. and SONTAKKE, N.A. 1990b. Surface and upper air temperatures over India in relation to monsoon rainfall, *Theor. Appl. Climatol.* **42**: 93-110.

PARTHASARATHY, B., SONTAKKE, N.A., MUNOT, A.A. and KOTHAWALE, D.R. 1990a. Vagaries of Indian monsoon rainfall and its relationships with regional/global circulations, *Mausam.* **41**: 301-308.

RAINBIRD, A.F. 1967. Methods of estimating areal average precipitation. WMO/IMD Report No. 3, WMO, Geneva, pp. 42.

RAMDAS, L.A. and MASWEKAR, S.L. 1922. Surface convection and variation of temperature near a hot surface. *Indian J. of Physics.* **7**: 1-13.

RAMDAS, L.A. 1946. The micro-climate of plant communities. *The Indian Ecologist.* **I(1)**: 1-22.

RAMDAS, L.A. 1951. Micro-climatology investigations in India. Archiv. Met. Geophys. *Bio. Met. Series* (B). **3**: 149-167.

RAMDAS, L.A. 1957. Natural and artificial modifications of microclimate. *Weather.* **12**(8): 237.

SCHICKEDANZ, P.T. and BOWEN, E.G. 1975. Computation of climatological power spectra using variable record lengths, Proc. of the Fourth Conference on *Probability and Statistics in Atmospheric Sciences* held at Tallahassee, Florida, USA, pp. 102-113.

SHEA DENNIS, J. and SONTAKKE, N.A. 1995. *The annual cycle of precipitation over the Indian subcontinent: Daily, Monthly and Seasonal Statistics.* NCAR Technical Note No. TN/401+STR, Boulder, CO, USA pp. 168.

SHUKLA, J. 1987. Inter-annual variability of monsoon, In: *Monsoons* (eds. J.S. Fein and P.L. Stephens), Chap. 14, Wiley and Sons, New York. pp. 399-464.

SINGH, N. 1994. Optimizing a network of raingauges over India to monitor summer rainfall variations, *Int. J. Climatol.* **14**: 61-70.

SINGH, N. 1986. On the duration of the rainy season over different parts of India, *Theor. Appl. Climatol.* **37**: 51-62.

SINGH, N. 1995. Large-scale inter-annual variation of the summer monsoon over India and its empirical prediction, *Proc. Ind. Acad. Sci.* (*Earth Planet Sci.*), **104**, 1-36.

SINGH, N. and SONTAKKE, N.A. 1999. On the variability and prediction of the rainfall of the post-monsoon season over India, *Int. J. Climatol.* **19**: 309-339.

SINGH, N. and SONTAKKE, N. 1996a. The instrumental period rainfall series of the Indian region: A documentation, *Research Report No. RR-067*, Indian Institute of Tropical Meteorology, Pune (India). pp. 79.

SINGH, N. and SONTAKKE, N.A. 1996b. Climate variability over Pakistan and its relationships to variations over the Indian region, In: *Climate Variability and Agriculture* (Y.P. Abrol, S. Gadgil and G.B. Pant eds.), Chap. 5, Narosa Publishing House, New Delhi (India). pp. 69-97.

SINGH, N., PANT, G.B. and MULYE, S.S. 1991. A statistical package for constructing representative area-averaged rainfall series and its updating using selected stations data, Proc. Natn. Semi. *Use Computers in Hydrology and Water Resources*, Indian Institute of Technology, Delhi16-18 December 1991, vol. II, pp. 122-134.

SONTAKKE, N.A. and Singh, N. 1996. Longest instrumental regional and all-India summer monsoon rainfall series using optimum observations: Reconstruction and update, *The Holocene*. **6**: 315-331.

TOMITA, T. and YASUNARI, T. 1996. Role of the northeast winter monsoon on the Biennial Oscillation of the ENSO/monsoon system, *J. Met. Soc. Jpn.* **74**: 399-413.

VERMA, R.K., SUBRAMANIAM, K. and DUGAM, S.S. 1985. Inter-annual and long-term variability of the summer monsoon and its possible link with northern hemisphere surface air temperature, *Proc. Indian Acad. Sci.* (*Earth Planet Sci.*) **94**: 187-198.

WALFORD, C. 1878. The Famines of the World: Past & present. *Journal of the Statistical Society*, Vol. XLI, part III.

WALKER, G.T. 1910. On the meteorological evidence for supposed changes of climate in India, *Indian Meteorological Memoirs*, **21**, Part I. pp. 1-21.

WALKER, G.T. 1914. A further study of relationship with Indian monsoon rainfall, *Indian Meteorological Memoirs*, **21**, Part VII. pp. 1-12.

WALKER, G.T. 1924. Correlation in seasonal variations of weather-IX: A further study of world weather, Memoirs of India Meteorological Department, New Delhi, **24**. pp. 275-332.

WEBSTER, P.J. and YANG, S. 1992. Monsoon and ENSO: Selectively interactive systems, *Q.J. Roy, Met. Soc.* **118**: 877-926.

WINSTANLEY, D. 1973. Rainfall pattern and general atmospheric circulation. *Nature*. **245**: 190-194.

WYRTKI, K. 1975. El Niño—the dynamic response of the equatorial pacific ocean to atmospheric forcing, *J. Phys. Oceanogr.* **5**: 572-584.

WYRTKI, K. 1982. The southern oscillation, ocean-atmosphere interaction and El Niño, *Mar. Tech. Soc.* J. **16**: 3-10.

WYRTKI, K. 1985. Water displacements in the Pacific and the genesis of El Niño cycles, *J. Geophy. Res.* **90**: 7129-7132.

YASUNARI, T. 1987. Global structure of the El Niño/Southern Oscillation. Part II. Time Evoluation, *J. Met. Soc. Japan.* **65**: 81-102.

Rice—in a Variable Climate, 1999, pp. 107-122
Eds. Y.P. Abrol and Sulochana Gadgil

7. Climate and Rice Production in India

Transformation of the age old subsistence farming into an intensive agriculture that enabled the country during the last three decades to march from an era of chronic deficiency and excessive dependence on import to a period of sustained self-sufficiency and surplus is the golden chapter in the history of post-independent India. The spectacular production advance has been possible because of four major research developments namely (i) exploitation of hybrid vigour in millets since late fifties, (ii) introduction and improvement of plant type-based high yielding varieties of wheat and rice since early sixties (iii) improvement of crop management technologies for sustainable productivity growth of all major commodity based cropping/farming systems and (iv) organized seed sector for production and supply of quality seed. Rice and wheat contribute today as high as 76 % to the national food grain production and of this rice alone accounts for 43 %. Unlike in wheat, where almost three fold increase in area (7 to 22 mha) has contributed substantially to its more than five time increase in production, in rice in the absence of appreciable increase in area, production advance is due to two fold increase in grain yield (< 1.0 to 1.9 tha^{-1}). Until 1980s production rise has been quite impressive. The annual growth rate was 4.7 % as against 2.8 % in the seventies with an overall mean growth of 2.9 % for 1967-68 to 1995-96, which is marginally higher than the minimum required level of around 2.8 % to sustain the present level of per capita availability (66 kg/year i.e. 30% of the caloric supply). The sustained high growth has been possible because of successively high yield growth in one region or the other. Whereas north west zone registered very high yield growth upto the end of 8th Plan, in south zone it was during 6th and 7th Plan periods and in west and eastern states since late eighties. Nevertheless what is disquieting is the sharply declined growth in the high productivity northern and southern zones and still very slow growth in eastern India, which together have brought down the overall growth to as low as 2.8 % (Table 1). More disturbing matter of concern is the further fall of rice production growth to 1.5 % by 1996-97 and consequent decline of food grain production growth to 1.7 %.

Table 1. Annual compound growth rate of rice production (mt) and yield (tha^{-1})

Zone	1982-1990		1991-1995	
	Production	Yield	Production	Yield
Northern	5.2	4.0	2.4	1.7
Southern	3.5	3.8	2.5	2.6
Eastern	6.8	5.6	2.1	2.1
Western	2.0	1.8	6.5	5.4
All India	4.7	4.0	2.5	2.8

To sustain the current level of foodgrain sufficiency demand, projection of rice is about 105 million tonnes by a decade from now. This requires a minimum annual yield growth of 2.4 %, with increase of the national average yield from the present 1.9 to 2.5 t/ha by 2006-07. With practically no scope for expansion of net area under rice and limited opportunities for bringing more area under irrigation, achievement of the production targets envisaged is not an easy task. Compelling socio-economic factors, especially scarcity of farm labour and increasingly prohibitive wage structure on the one hand and decline net return-necessitated crop diversification on the other have brought down *kharif* rice area by 1.5 million ha and after deducting one million ha increased under *boro* (winter) rice, the net decline during the 8th Plan period is around 0.5 million ha. Plateauing of yield level in all the major zones, no appreciable genetic yield improvement since the introduction of dwarf varieties and declining factor productivity in the major rice-based cropping systems, especially rice-wheat in the Indo-Gangetic Plains (10 mha) and rice-rice in the traditional rice belt further compound the problem making it all the more difficult and most challenging.

One other factor, which determines year to year prospects of agricultural production in general and rice in particular is the climate-related physical and biological constraints. Among the crop plants the most vulnerable to environmental influence is rice. It is a crop grown under extremely diverse agroclimatic conditions - from below mean sea level in the Kuttanad region of Kerala to as high an altitude as>8000 feet in the Himalayan mid-hill and extreme moisture regimes ranging from low rainfall (<400 mm) drought-prone uplands to high rainfall (>3000 mm) deep water areas of eastern India (Table 2).

Table 2. Rainfall in the eight agroclimatic regions of India.

Sl.No.	Region	Rainfall Range (cm)	
1.	Humid Western Himalayan Region (J&K, HP and Hills of UP)	8 (Ladakh)	350 (Hill of HP & UP)
2.	Humid Bengal-Assam Basin	220	400
3.	Humid Eastern Himalayan Region (N. E. States and Andaman Nicobar Bay Islands)	200	400
4.	Sub Humid Sutlej-Ganga Alluvial Plains (Punjab and Plains of UP, Bihar)	30	200
5.	Sub Humid to Humid Eastern & South Eastern Uplands (Orissa, AP and Eastern M. P.)	100	180
6.	Arid Western Plains (Haryana, Rajasthan and Gujarat)	10 (Extreme West)	65 (Extreme East)
7.	Semi Arid Lava Plateau & Central High lands (Maharashtra, West & Central MP, and Goa)	70	125
8.	Humid to Semi Arid Western Ghats and Karnataka Plateau (Karnataka, TN, Kerala, Pondicherry, and Lakhshadeep).	60 (Plains)	300 (W. Ghats)

Source : Velayutham, *et al.* 1996.

India's rice growing areas have been grouped broadly into three climate regions based on monsoon pattern by Huke (1976) (Fig. 1). Region 1 comprising Ganges Delta, Assam, West Bengal, eastern part of Bihar and western coastal lowlands is characterised by heavy rains enough to provide only modest shifts in output, however, low it is. Here nature alone controls all the climate parameters required by the crop. The climate although is neither perfect nor optimum, it is safe. Central part of the country extending from U.P. to western Bihar in the north and Tamil Nadu in the south consititute the Region-2, the most important rice producing area. In this region atleast one parameter of the climate is frequently less than the ideal for a successful crop, which man can control effectively. Sizable portion of it is either rainfed or irrigated and irrigation water is largely from rain- fall-dependent small tanks and shallow wells. Mean total rainfall is marginal for rice and year to year variability in terms of time of monsoon arrival and total volume of precipitation is quite wide. High energy investment is its limitation. Region 3 comprising north western part of India is characterised by very low and year to year highly variable rainfall. Nevertheless, the climate is modest for rice, which is predominantly irrigated.

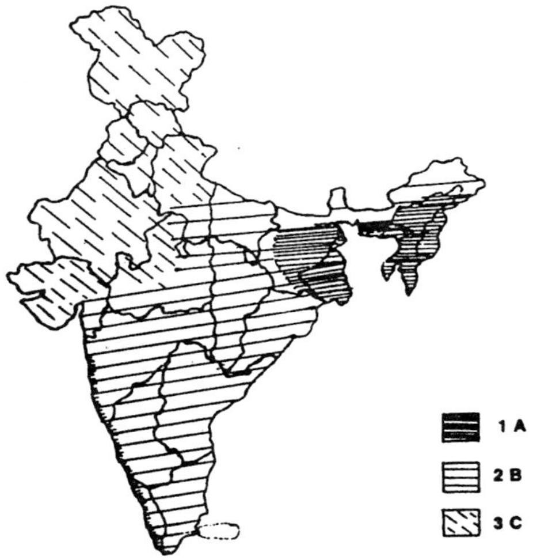

Fig. 1: Climate and rice. (1) "Home" area of rice where the crop may be raised year after year with no climate modification by man. (2) Important rice producing areas where at least one parameter of climate is frequently less than the ideal for successful crops. (3) Areas where rice production is widely scattered and where climate must be modified to produce a crop. Based on Huke (1976).

Across the agroecological regions, components of the rice growing environment, especially monsoon behaviour show no major change except for year to year fluctuation in the quantum and distribution pattern. Analysis of 120 year (1870 to 1990) data on monsoon behaviour reveals that the total rainfall averaged over the long period is around 88 cm, good monsoons come in sequence of upto 20 years and deficit of rainfall barring exceptional drought years is between 5-15 % less than the normal. Even in the lowest rainfall state of Rajasthan, no major shift in the monsoon behaviour has been experienced during the last 65 years. The country has been enjoying favourable monsoon, consecutively for the 10th year with 80% of India's districts (33 to 35 meterological sub-divisions) receiving rainfall in the range 91 to 119 % of normal (Fig. 2 & 3). This is not unusual as once for over 20 years (1921-1940) and another time for 13 years (1952-65) monsoon was consecutively normal. Whatever excess or deficit has been experienced during the last 10 years is due to erratic volume and distribution pattern as summarised in Table 3. This does not mean that the climate is same as it was 1,000 years back. Changes might have been within the broad natural climate cycle, which is influenced by cooling or warming of the globe.

Table 3. Crop losses as affected by monsoon behaviour during 1990s.

Monsoon Behaviour	Year	Eastern India	Rest of India
Delay in monsoon & delayed planting	1991	—	UP, HP, Haryana
	1995	—	AP (Telengana)
Deficit rain & drought spells	1992	Bihar (North)	AP (Telengana)
	1994	Assam, Bihar (North)	AP (Telengana)
	1995	Orissa, MP	Maharashtra, Gujarat(North), Parts of UP, AP, TN, Karnataka
Excess/Heavy rains during flowering/ grainfilling	1991	—	AP (Coastal), Kerala
	1992	—	Kerala
	1995	—	AP(Coastal), Punjab, Haryana, Rajasthan
Cyclones & floods	1991	—	AP(Coastal), Kerala
	1992	—	HP (Hailstorm)
	1995	Bihar (North)	

The period between 1890 and 1945 characterised by rising temperature globally, is considered to be the most favourable part of long term natural cycle for agricultural productivity. It is largely the differences in temperature that change the course of wind systems, which in turn determine the quantum and distribution pattern of precipitation resulting in drought or floods in one part or the other of the World. In the Indian sub-continent, as critically analysed by Bryson (1974), recurrence of drought for instance, varied from period to period. Before 1920, north western India experienced years with less than half of normal rainfall, the probability of drought visiting after every 6-8 years. Between 1920 and 1960, the return period rose to 14 years and during 1930s once in 18 years. Similarly period to period variation in excess rain causing floods is also not uncommon. During the last 40 years, area affected and crops lost due to floods are on the increase (Table 4).

Table 4. Annual average flood affected areas in India*

Period	Area affected (mil. ha)	
	Total	Cropped
1950s	6.86	2.08
1960s	5.86	2.47
1970s	11.19	5.55
1980s	16.57	6.91

* Source : Centre for Science & Environment, New Delhi (1991)

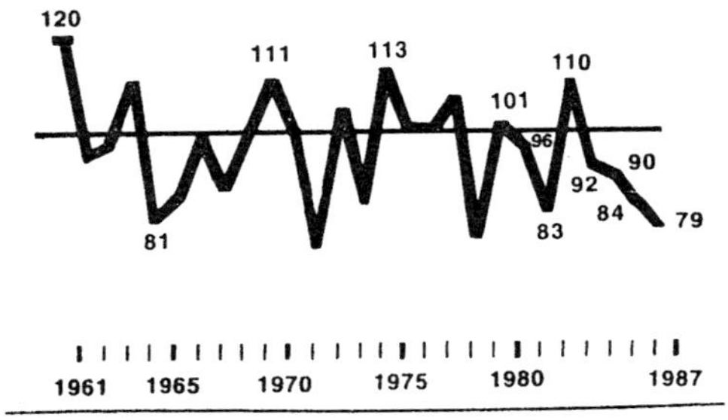

Fig. 2: Trend of rainfall during sw monsoon (as percentage of normal)

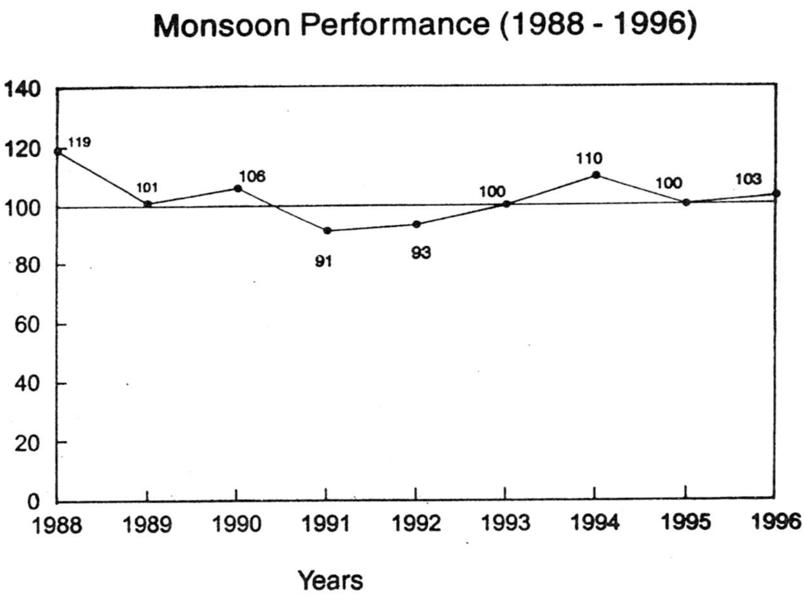

Fig. 3: Actual rainfall as percentage of normal rainfall.

7.1. Rainfed Ecologies and Characterization of Monsoon Period *vis-a-vis* Crop Growth Stage

As eastern-north eastern India represents the typical rice growing humid tropical region with all major weather-related constraints affecting rice production/productivity, the following discussion on the influence of climate is confined to this predominantly rainfed region.

The region comprising Assam, Bihar, Orissa, West Bengal, Eastern U.P., eastern M.P. and the north eastern hill states accounts for 60 % of the total rice area and 48 % for the national production. with an average yield of 1.3 t/ha. More than 80 % of the area is rainfed, the rainfall ranging from <1000 mm to >3000 mm (Fig. 4). The rainfed ecology consists of diverse water regimes viz lowland comprising 48 % (shallow upto 30 cm water depth) 40 % and semi deep water upto 70 cm depth 8 %, 5 % deep water (>1 meter depth) and 16 % upland. Constraints to the crop productivity, which vary with the water regime are predominantly climate-related viz. volume and distribution pattern of rainfall, minimum-maximum temperature, solar radiation and weather influenced disease pest incidence.

Moisture availability, which depends on the relative volume of precipitation at different phases of monsoon season *vis-a-vis* crop growth stage determines the crop prospects. Moisture availability period as measured by balance between rainfall and potential evapotranspiration (PET) has been divided into three sub-classes viz. Moist I, Humid and Moist II by Singh et al., (1996) (Fig. 5). Moist I and Moist II have been defined by them as the period, when the volume of rainfall is less than PET but more than 1/2 PET. Moist I characterised by relatively high temperature and longer day length, besides high PET is at the beginning of the monsoon and most of the crop establishment operations are carried out at this phase. Moist II, characterized by low PET, low temperature and shorter day length is at the end of the season, when crop harvest and post-harvest processing operations take place. Humid is moisture stress-free period, when rainfall is in excess of PET and crop completes nearly its vegetative growth and enters reproductive phase. According to them moisture availability period varies with the location. For instance, the duration of Moist I and Humid is relatively shorter in U.P., M. P. and Bihar, while it is longer in Orissa, West Bengal, Assam and North-eastern hill states. The duration of Moist II, however, is longer in U.P., M. P. and Bihar but shorter in the other three states. Their finding further reveal that since crop establishment for varieties of different duration is nearly fixed at a certain point of time, long duration varieties suffer more from water stress at reproductive and maturity stages as compared to medium and short duration varieties. Such information would be of value in preparing crop production strategies and cropping calenders as well as in delineating the technology application domains.

7.2. Weather-Related Constraints

Entire rainfed rice area in the eastern India is prone to one or the other of the weather-related stresses such as drought, water logging/flash flood, low solar radiation and disease-pest pressure. The stresses occur to different degrees at all stages of crop growth but vulnerable phases differ depending on the ecology. In the upland, for instance, drought affects the crop at the initial and terminal stages of growth depending on the time of sowing and crop duration. Effect of inadequate and erratic rainfall in this ecology is aggravated by the poor water holding capacity of its light soils as well. Drought at seedling and flowering stages together has been estimated to account

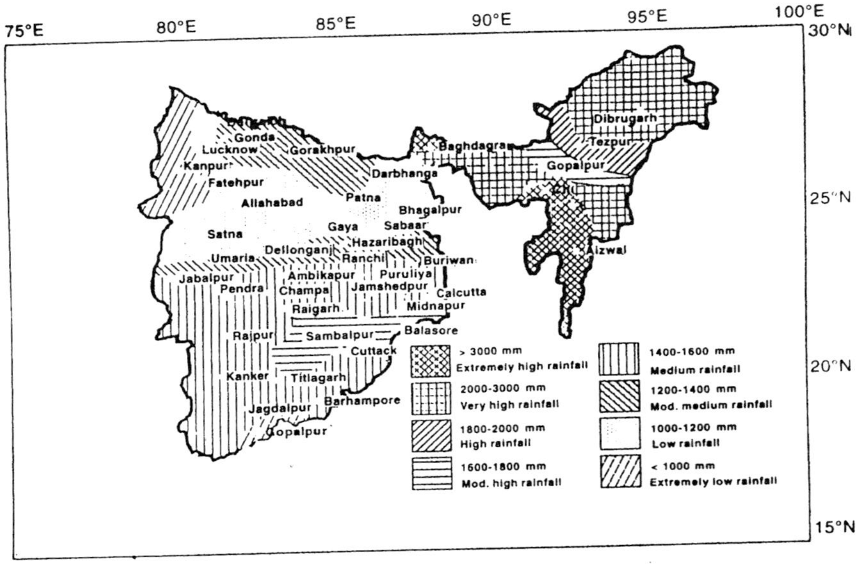

Fig. 4: Annual rainfall in Eastern India.

for an annual loss of three million tonnes in this region alone. In lowland and deep water ecosystems submergence is the most critical constraint to rice production. Majority of the lowland varieties can neither elongate to cope with rapid rise of water level nor survive inundation. Even under temporary inundation or flash floods, they suffer significant yield losses. Floods occur right from the month of

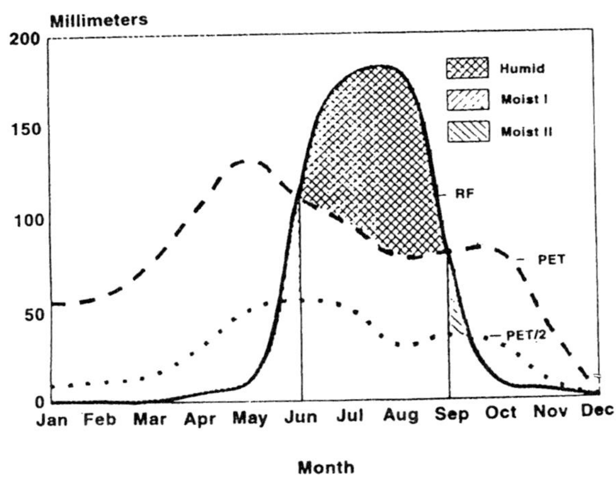

Fig. 5: Moisture availability analysis with respect to rainfall and potential evapo-transpiration (PET).

June, when the crop is too young, to September, when it is at flowering/grain filling stage. In as many as six districts (Lakhimpur, Sibsagar, Dibrugarh, Darang, Naogong) of Assam on an average 2,50,000 ha of rice area are affected by floods every year. In Orissa, about 3,80,000 ha are vulnerable to floods and similarly in West Bengal (Midnapur, Howrah, Hoogly, 24 Parganas, Nadia, Murshidabad and Malda) over 4,00,000 ha are prone to floods and deep water. Flash floods and waterlogging are the chronic problems in several districts, (Nalanda, Samastipur, Darbhanga, Saran, Siwan, Patan, Bhogpur, Rohtas, Gopalganj, Saharsa, Modhepura, Purnea, Kathuhar and Khagaria) of Bihar.

Analysis of drought and flooding pattern in the shallow lowland ecosystem, reveals that 54.6 % of the area (5.7 mha) under this ecology is drought prone, 25.5 % (2.7 mha) drought and submergence prone, 10.2 % (1.1 mill. ha) submergence prone and hardly 9.6 % (1.0 mha) is favourable. The entire area in the medium deep rainfed lowlands is submergence prone.

In the tropics in general and eastern India in particular, low solar radiation (short sunshine hours due to cloudy weather) rather than temperature, is the major constraint besides deficit/excess of moisture to the productivity of late maturing rices (*sali* in Assam and *aman* in West Bengal). Some is true even in the west and east coastal districts of peninsular India. The intensity and period of sunshine varies with the location, month and crop growth stage. Lowest solar radiation (<200 to >300 cal/cm^2/day) is in the months of July and August, when the crop is beyond maximum tillering stage, while it is in the range of <300 to >400 cal/cm^2/day during October-December, when the crop is in flowering/grain filling stage in this region as against 400 to 600 cal/cm^2/day in other regions. Low radiation at flowering-grain filling stage invariably results in low yields as it impairs seriously spikelet fertility.

Since the advent of high yielding varieties there has been a shift in the spectrum and intensity of pests and pathogens across ecologies (Fig. 6 & 7). Whereas we knew no more than a couple of fungal diseases (blast and brownspot) and insect pests (stem borer and gall midge) prior to high yield technology, the spectrum has grown so wide today that there are additionally four more fungal (sheath blight, sheath rot, rice bunt, leaf scald) two bacterial (bacterial leaf blight and bacterial streak) and three viral (Rice tungro virus, grassy stunt virus I and II and ragged stunt) diseases and many new insect pests and their biotypes (5 biotypes of gall midge, brown plant hopper, green leaf hopper, white-backed plant hopper, leaf folder, hispa, ufra nematode) have been added to the short list of the early sixties. Genetic homogenity, conductive crop canopy under conditions of high input management, staggered planting/cropping intensity now pos-

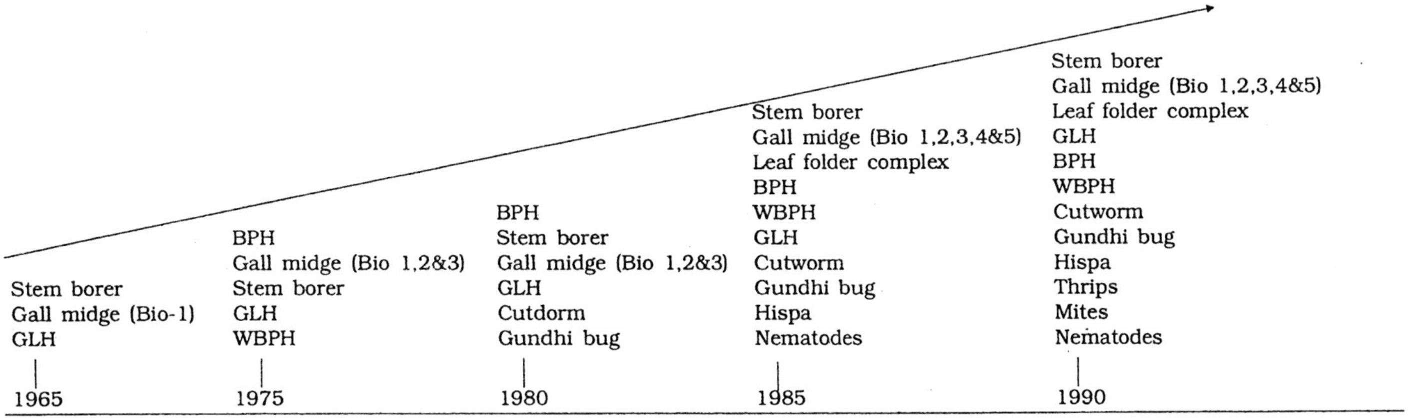

Fig 6. Changing Pest Scenario and emergence of new pests and biotypes

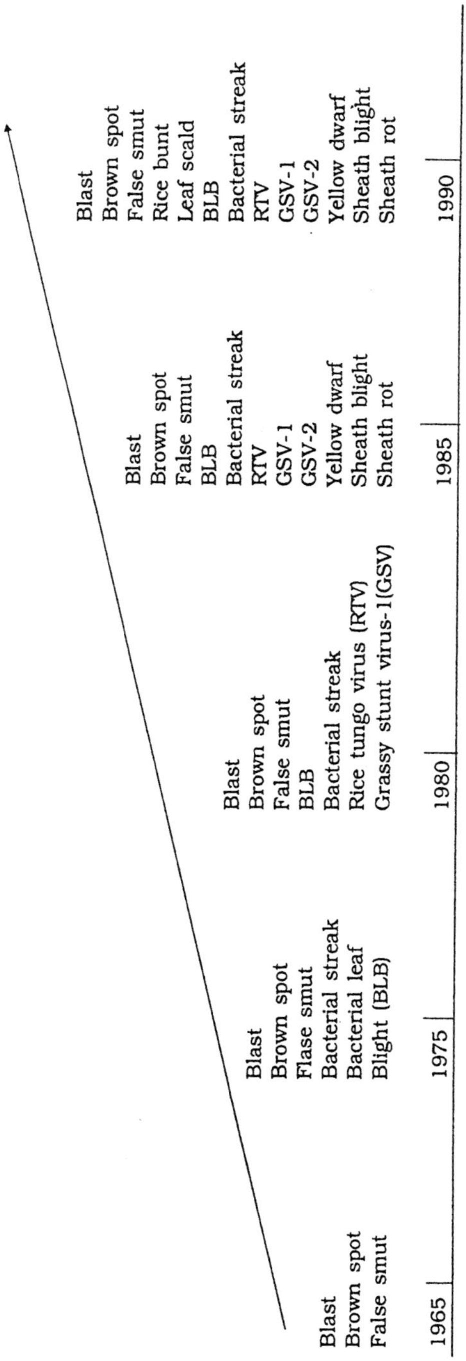

Fig 7: Changing disease scenario and emergence of new diseases.

sible because of earliness photo-insensitivity of the high yielding varieties, could be no doubt, the reasons for the increasingly complex disease-pest syndrome. We, at the same time, have no reason as yet to explain equally high incidence of diseases and insect pests in the native varieties as well. Increasing temperature, possibly due to increasing load of greenhouse gases as well as CFC depleted ozone layer resulting in increased UV-B radiation could as well induce changes in the epidemiology and virulence/viruliference of the pests. Also, there is the likelihood of once unimportant organisms assuming serious pest proportions. More in depth anticipatory research on the pest syndrome is, therefore, urgent and relevant so that we are ready with matching technology to face such eventualities.

Silent change in the climate is undoubtedly a reality notwithstanding the differences of opinion on its pace and likely impact on agricultural productivity. There is, therefore, an urgent need now than at any point of time before to assess precisely the rate of change taking place in all the major parameters of climate and to reorient accordingly the research and development programmes to cope with their adverse/favourable impact. The following are some of the areas that warrant our attention, (a) characterization and demarcation of areas on the basis of their relative vulnerability to weather aberration (b) perfection and use of weather forecasting system employing all available state of art technologies to plan cropping strategies (c) accelerated research to develop appropriate crops/varieties to utilize still unexploited and underexploited seasonal windows (d) identification and use of both natural and novel gene sources for adjusting to the anticipated variations in the different parameters of climate change. Increasing level of CO_2 and other greenhouse gases resulting in the rise of atmospheric temperature consequent effect on sea level leading to inundation of coastal areas and enhanced UV-B radiation causing changes in the pest-disease syndrome are some of the anticipated changes in the biosphere. Increased temperature will be beneficial to cooler regions while it is the deleterious in the warmer regions. Also, the scanty rainfall areas would become still drier while others will become wetter. By and large the impact of climate change on rice culture will be more and more complex. Development of precise climate models can help anticipate the nature of change or mitigation of adverse effect of such change (See chapter 8).

It is increasingly evident that excessive dependence on the single strategy of 'varietal solution' to the progressively complex and diverse rainfed ecologies, is netiher wise nor would be relevant in the coming decades. What is required is development and adoption of tactical farming approaches to minimise crop losses through risk-distribution strategies in years of unfavourable weather and maximise the productivity potential in years of favourable weather. In both the cases, The variety should be a high yielding genotype with high level of resilience to adapt to one or two key components of the stress environment. The following are some of the widely adopted practices developed on the basis of centuries of experience and wisdom of farmers in the tropical humid parts of India. Scaling up of such practice through replacement of native varieties with modern high yielding varieties and improved production skills is the right short term strategy to enhance and stabilize rice productivity of areas in uncertain monsoon behaviour.

7.3. Flexibility in Crop Choice and Cropping Patterns

In areas vulnerable to cyclones, sea water intrusions, recurrent floods etc. not only the rice crop suffers, but more so the entire rice based cropping pattern *in vogue* there. In such situations, possibilities, wherever feasible with least compromise on ecological and economic security, of finding alternate cropping patterns should be explored. The strategy could be either redesigning of the system to maximise production outside the disaster season or substituting rice with an alternate crop, which is not vulnerable to hostile weather.

7.4. Cropping Patterns for Risk Distribution

In low lying areas as in Assam, rice crop is damaged almost regularly by floods occurring between May and August. The floods generally damage the *pre-kharif* '*ahu*' (March-June) as well as the winter/*kharif* '*sali*' rices (July-November). Sometimes deep water '*bao*' (March-November) rice also becomes vulnerable to floods as late as September end. To distribute the risk, farmers practice mixed planting of early maturing *ahu* varieties with late maturing deep water (*Bao*) rices in the flood prone areas. If floods occur during the early part of the season short duration '*ahu*' crop is damaged leaving the late maturing *kharif* (*bao*) rice to survive and yield atleast 60% of the total expected. If floods occur late in September/October, the *bao* crop is lost, while a good crop of '*ahu*' had already been harvested. In years of 'no floods' '*ahu*' and '*bao*' crops raised in 1:1 ratio, yield together very high.

Besides rice-rice there is scope for rice-non rice pattern as well. In the deep water areas of Bihar, the practice for centuries is to take only one crop of rice. To increase the productivity of such areas mixed/inter cropping of short duration non-rice crops like *mung* bean, forage sorghum, seasame etc. could be grown successfully with long season rice. The companion crop planted in February-March is harvested by June leaving the rice crop. Incidentally this strategy is already being practiced by enterprising farmers in the region. In the drought and submergence prone lowland sub-ecology as in eastern U.P., deep water rice is grown with maize as mixed or inter crop. In years of drought, maize proves a 'saviour crop' to the farmer and in years of excessive rainfall and resultant inundation to varied depths, rice flourishes and meets the farmer's requirements. In the rainfed upland areas such as Adilabad and Agency areas of Andhra Pradesh growing rice in alternate rows with maize is a common practice for taking a 'sure crop'. When rainfall is normal and above, rice thrives well and yields handsomely and in years of severe drought, deep-rooted pigeonpea survives and yields well, when rice crop dries up.

7.5. Land Shaping and Harvesting of Rain Water

In the coastal areas of West Bengal, farmers have to cope with erratic rainfall, soil and ground water salinity, lack of irrigation and drainage facilities and occasional floods. Based on their native wisdom a simple technique has been devised to overcome the stress effect and thereby improve the productivity of the land. The technique consists of digging of a tank of about 2.5 meter depth in about one fifths of the field area for storage of rain and flood water. The level of the rest of the field is raised by 30-50 cm using the dug out soil. This strategy helps reduce waterlogging to an extent on the one hand and makes available the raised field for growing a second/third non-rice crop immediately after the wet season rice crop on the other. Water stored in the field tank is used for the second/third crop as well as for fish culture. (Mandal et at., 1990)

7.6. Tactical Production Strategies

Changing the crop calendar : To avoid cold injury to the '*boro*' crop during December-January as well as to save the costly irrigation water, especially during the early part of the season, a combined crop of '*boro*' and '*aus*' called '*Baraus*' (February to June) has been designed and being practiced in the adjoining Bangladesh. While ensuring high yield, the practice allows timely planting of '*aman*' rice. (Bhuyian and Galong, 1987)

Timely establishment of wet season crop : Either delayed arrival or weak monsoon invariably exposes the crop to moisture stress at reproductive phase. Waiting for active monsoon under such circumstances delays timely transplanting and crop establishment resulting in low yields. In eastern M.P., Bihar and Orissa, where delayed onset of monsoon is almost a regular feature, dry-wet cultivation called '*Beausani*' in place of transplanting or dry direct seeding is the practice to ensure timely planting. The system of '*Beausani*' consists of dry sowing immediately after receiving pre-monsoon showers and cross ploughing/laddering, when the fields get flooded. This has the advantage of timely planting and full/effective utilization of total precipitation received and thereby avoiding/minimising the crop losses from intermittent spells. Also, this is an effective practice for weed and insect pest (gall midge) control. An improved version of the system developed by IRRI is worth trying for extensive adoption. The system consisting of line sowing instead of broadcasting and intercropping with the green manure crop, *Sesbania rostrata* ensure timely planting, optimum plant population, mechanical weeding and nitrogen supply from the green manure. crop incorporated after a month and a half, and it has been found more productive than the traditional '*Beausani*'

7.7. Choice of Cultivars

In the areas of uncertain climate and hydrological conditions, apart from the above said tactical crop management practice, adoption of appropriate varieties would help enhance the productivity. Desirable traits to ensure yield security in a variety should include (a) seedling adaptability in the nursery to delayed transplanting with least effect on yield (b) adaptability of young plants to submergence for relatively longer period at times of uncontrollable rise of water depth (c) plasticity to tiller over relatively longer periods as a means of recovery from early drought damage (d) tolerance to drought, low temperature and poor light (low solar radiation) at critical stages of crop growth (e) photosensitivity or insensitivity respectively in long and short season varieties and (f) desired level of resistance to key pests. Screening large volumes of native and exotic germplasm under artificial and field conditions at various rice research centres/hot spots has led to the identification valuable donor sources for most of the traits. While breeding efforts are accelerated for evolving gene sources, continued collection and conservation of germplasm should receive special attention. Any slackness in this matter would hasten the replacement of thousands of land races in the weather-related stress environment by a few high yielding varieties resulting in irrecoverable loss of invaluable genes, the nature and our ancestors had nurtured for centuries. In the event of exhaustion or non-existence of desired candidate genes in the primary and secondary gene pools innovative biotechnology-genetic engineering tools be employed to scout and use heterologous genes from distantly related species and unrelated organisms.

7.8. References

BHUIAN, S.I., GALANG, A.L. 1987. Impact of climatic factors on irrigated and partially irrigated rice cultures. In: *Weather and Rice*, Proc. Workshop, IRRI, Los Banos, Philippines, April 7-10, 1986. p. 121-126.

BRYSON, R.A. 1974. A perspective on climate change. *Science*. **184**: 753-760.

MANDAL, B., DAS, S.C., BHANDARI, B. 1990. Sustainable rice breeding farming systems in rainfed low lands of coastal West Bengal. In: (Thuvalidhara *et al.* eds), *New Frontiers in Rice Research*. Proc. Int. Symposium, DRR, Hyderabad, Nov. 15-18, 1990. p. 292-293.

HUKE, R. 1976. Geography and Climate of Rice. In: *Climate and Rice*, IRRI, Manila, Philippines. p. 31-50.

SINGH, V.P., SASTRI. A.S.R.A.S., BAGHEL, S., CHAUDHARY, J.L. 1996. Rice growing: Environmental Analysis of Eastern India—An Agroclimatic Atlas. Vol. I, IRRI, Manila, Philippines/Indira Gandhi Agricultural University, Raipur, India. pp. 1-96.

VELLAYUTHAM, M., RANA, K.P.C., SIDHU, G.S., KANDPAL, B.K., LAL, T., SINGH, S.P., BHATTACHARYA, T. 1997. Characterisation of agro-eco sub regions in the Indo-Gangetic alluvial plains for sustainability of rice wheat cropping systems. Paper presented at Regional Co-ordination Commitee for South-East Asian countries, Pokhra, Nepal, June 25-26.

AGRAWAL, A., NARAIN, S. 1997. *Flood, Flood plains and Environmental myths.* Centre for Science and Environment, pp. 167.

Rice—in a Variable Climate, 1999, pp. 123-150
Eds. Y.P. Abrol and Sulochana Gadgil

8. Growth and Yield Responses to Weather Components and Predicted Climate Change

Rice is cultivated in diverse ecologies which differ from each other in water availability and depth of standing water during growth. In India, the rice crop is grown in about 42 million ha. At present, the area of irrigated ecosystem is about 19 million ha with a productivity of 3.6 t/ha. The rainfed lowland area is about 14 million ha with a productivity of 2.4 t/ha. Ninety percent of the total production of rice is contributed by these two ecosystems. The upland ecosystem is nearly 6.5 million ha with a low productivity of only 0.8 t/ha. About 3 million ha is flood-prone for a variable period during growth and, based on the depth of water, it is described as semi-deep (50-100 cm) and deep water (more than 100 cm). The average productivity of flood prone ecology is about 1.5 t/ha. About 50 per cent of total rice production in India comes from 4 states (West Bengal, Andhra Pradesh, Uttar Pradesh and Punjab). Bihar, Orissa, Tamil Nadu and Madhya Pradesh contribute about 33 per cent of total production.

Most of the area under irrigated ecology is in the southern and north-western parts of India. The rainfed ecologies (flood prone, lowland and upland) are mostly in eastern India. More than 90 % of rice is grown in the wet (monsoon) season (June-October) throughout the country. In hilly regions of Uttar Pradesh, Himachal Pradesh and Jammu & Kashmir, cold adapted rice varieties are grown during spring and monsoon seasons. In southern and eastern India, rice is grown in about 3.5 million ha during dry season (December-April) also in areas where water is available for irrigation. Fig. 1 shows rice growing seasons in different parts of India.

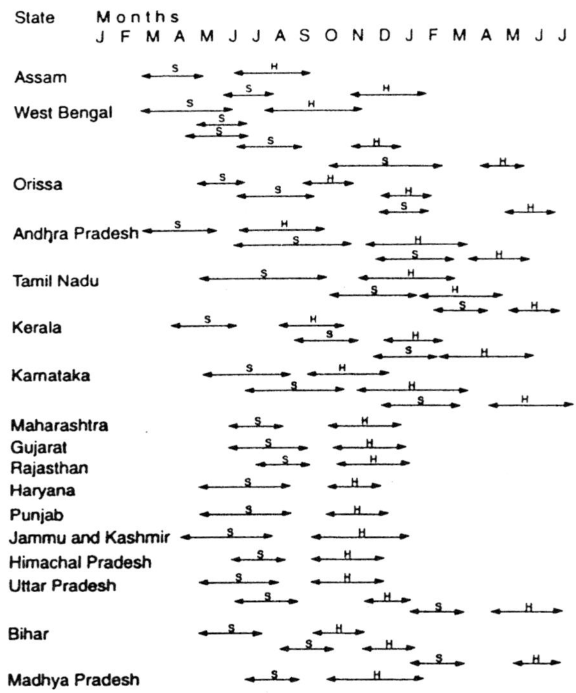

Fig. 1: Rice growing seasons in different parts of India (S: Sowing; H: Harvest) (Sreenivasan, 1980)

Weather variables influence not only production of dry matter during growth and development but also determine yield components in a season. Increasing CO_2 concentration in atmosphere and the predicted global warming would influence growth and productivity of rice in future climate change.

8.1. Growth and Developmental Phases

Rice crop takes about 3 to 6 months from germination to harvest, depending upon genotypes and environment (Vergara 1970; Yoshida

1981). Botanically, rice has two sequential growth stages; vegetative phase from germination to panicle initiation and reproductive phase from panicle initiation to maturity. But conventionally in rice crop

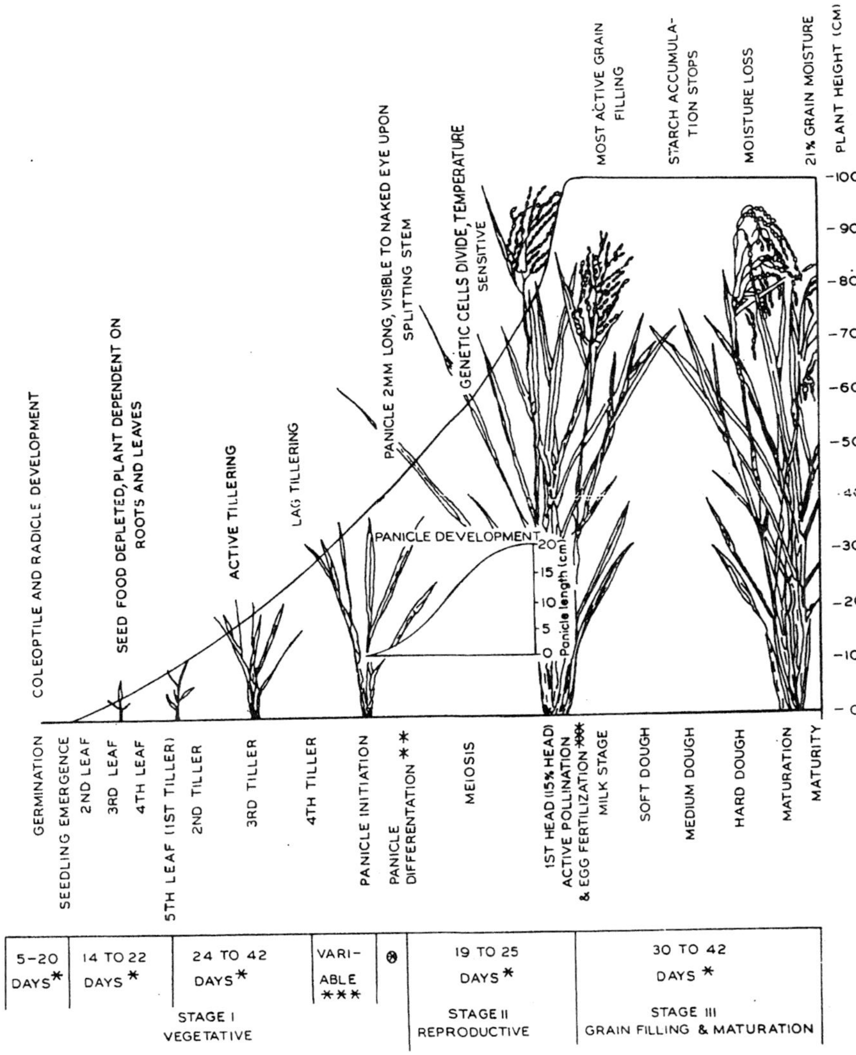

Fig. 2: Development stages of the rice plant. *3-5 days. Under warm conditions use the lower number of days and for cool conditions use the higher number of days. **The reproductive stage begins with panicle differentiation, which can be seen only by splitting the stem lengthwise. At this stage 30 % of the main culms sampled have a panicle 2 mm or longer. ***Stage III begins when 50 % of the florets pollinated. ****Variable time—0-25 days (depending upon variety).

physiology, the reproductive phase is considered to be from panicle initiation to flowering and the ripening or grain filling phase is from flowering to maturity. The duration of a rice genotype depends on duration of the vegetative phase as the duration of reproductive and grain filing phases is about 60 days in all genotypes.

In vegetative phase, leaves emerge at regular interval and tillering begins at 5th or 6th leaf stage. The vegetative phase comprises two component-phases: the basic or active vegetative phase (BVP) and photoperiod-sensitive phase (PSP) or lag-vegetative phase. BVP is the minimum period of vegetative phase before plant will initiate panicle primordium. Variations of daylength have no effect on its duration. Low temperature lengthens and high temperature shortens its duration. For most varieties, BVP is from 25 to 65 days. The optimum photoperiod of most varieties is about 9-10 hours. Duration of PSP varies greatly according to daylength in sensitive varieties.

In reproductive phase, panicle primordia initiation starts usually about 30 days before flowering, culm internode elongates increasing plant height, last leaf (flag leaf) emerges, panicle exerts and flowering (antesis) occurs. Within a panicle, most spikelets complete anthesis in about 5 days beginning from apex but it takes 10-14 days for flowering to complete in a field due to variation in panicle exertion within tillers of a plant and variation between plants.

In the grain filling phase, rice grains develop after pollination and fertilization. Based on texture and colour of growing grains, the ripening phase is termed milky, dough and maturity stages. The duration of grain filling is from 25 to 35 days.

Rice yield capacity is determined during vegetative (number of panicles per unit area) and reproductive (number of spikelets per panicle) phases. Actual yield is realised at flowering and during grain filling (filled spikelet % and weight per grain) phases (See figure 2).

8.2. Growth and Yield Responses to Weather Components

The effects of weather components on rice physiological processes and growth have been well studied. Temperature and solar radiation (intensity and duration) are the two important weather components determining yields in all rice growing ecologies. In addition, rainfall (amount and distribution) in a season is a direct critical weather variable in rainfed ecologies (lowland, upland and flood prone).

8.2.1 Temperature

In wet season, the maximum air temperature is around 30°C during most of growth period but crop may be subjected to relatively high temperature, if there is a break in or early withdrawal of monsoon. In the dry season, high temperature prevails at flowering and during grain filling phase. Low temperature is also a climate limitation in the early part (December to February) or dry season in ėastern India.

Low temperature limits rice growth and yield at higher latitudes (sub-Himalayan rice growing areas) during seedling and vegetative phase in spring season and during grain filling phase in wet season.

Temperature influences growth rate, duration and productivity. Generally, growth rate increased linearly in the temperature range of 22 – 31°C and higher temperatures adversely affected rice growth and productivity (Yoshida, 1981). A rice crop requires about 2000-4000 degree days which corresponds to 80-160 days when grown at a mean temperature of 25°C. A generalized relationship between temperature and length of time required to complete development is curvilinear, indicating that time required for plant development is lengthened below and above optimum temperatures. Optimal temperatures for various physiological processes/growth stages are presented in Table 1.

Table 1. Optimum temperature for different physiological processes or growth stages of rice

Physiological processes/growth stages	Temperature (°C)
1. Cell division	25
2. Cell elongation	30
3. Photosynthesis	20–35
4. Respiration	20–32
5. Translocation	18–29
6. Germination	20–35
7. Seedling emergence and establishment	25–30
8. Rooting	25–28
9. Leaf and stem elongation	22–31
10. Tillering	25–31
11. Initiation of panicle primordia	25–30
12. Panicle differentiation	18–30
13. Anthesis	30
14. Ripening	20–25

From: Nishiyama (1976), Venkataraman (1987), and Yoshida (1981)

Optimal temperature for ripening is lower than that for tillering and anthesis. Obviously, temperature optimum shifts to relatively lower temperature as rice grows.

Top growth of rice plants after transplanting is in general accelerated linearly from 18 to 33°C and the growth notably decreased above or below this range (Yoshida 1973). Rooting of rice seedlings occurred favourably over a range from 19 to 33°C with an optimum at 25 – 28°C (Chamura and Honma, 1973). Tillering increased with rising temperature in the range of 15 to 33°C (Yoshida, 1973). The optimal temperature for tillering was found to be 25°C at day and 20°C at night (Sato, 1972). Temperature above 33°C did not favour tillering (Chaudhary and Ghildyal, 1970). Panicle differentiation occurred in the range of 18 to 30°C (Adachi and Inoue, 1972). High temperature generally accelerated floral initiation (Vergara *et al.* 1972). At flowering and during grain filling phase, high temperature reduced yield by causing spikelet sterility and shortening the duration of grain filling phase. The fertility of spilkelets decreased as temperature increases above 30°C (Fig. 3). High temperature desiccated pollens (Osada *et al.* 1973). At high temperature of 35°C (day) and 30°C (night), sterility increased as a result of smaller pollens and non-dehiscence of anthers (Sato *et al.* 1973). Even one or two hours of high temperature at anthesis resulted in very high sterility (Satake and Yoshida, 1978). High air temperatures above 30°C were not favourable for ripening (Osada *et al.* 1973). Length of ripening period is inversely correlated with daily mean temperature and, therefore, grain filling is poor when temperature is above optimum, although rise in temperature increases rate of grain filling.

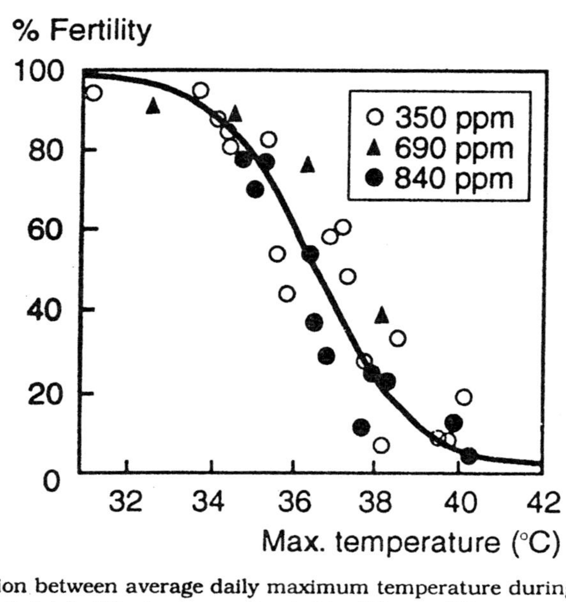

Fig. 3: Relation between average daily maximum temperature during the flowering period and spikelet fertility in rice acclimated to different CO_2 concentrations (Horie, 1993).

Water temperature affects yield by influencing panicle number per plant and spikelet number per panicle because growing points of leaves, tillers and panicles are under water at early growth stage under irrigated and lowland rainfed conditions. At later stage, air temperature controls panicle growth and ripening.

Low temperature effects are poor germination, delayed seedling emergence, stunted growth, leaf discolouration, panicle tip degeneration, incomplete panicle exertion, delayed flowering, failure of dehiscence and fertilization, high spikelet sterility and irregular maturity (Chang and Oka, 1976; Yoshida, 1981). Stunted seedling growth could be commonly seen in *indica* varieties under low temperature (Kaneda and Beachell, 1974). The lower limit for leaf elongation was 12°C. (Tanaka and Munakata, 1974) and 15°C for floral initiation (Owen, 1972), panicle development and exertion (Carnahan *et al.* 1972). The critical cool temperature causing sterility was 15 ± 5°C (Sasaki *et al.* 1973). Rice was most sensitive to low temperature at 2 to 1 week before flowering and at flowering (Sasaki and Wada, 1973). Cool night temperature below 15°C for 10-14 days before heading caused high sterility of spikelets. Both day and night temperature influenced spikelet sterility. Low temperature caused excessive shattering of grains. In general, *indica* varieties are more susceptible to low temperature then *japonica* varieties.

8.2.2. Solar Radiation

Solar radiation (intensity and duration) is a critical weather component during wet (monsoon) season, because of cloud cover during most of crop growth period. In the dry season, solar radiation is usually at optimum level during crop growth from transplanting. Solar radiation intercepted in rice canopy plays a major role in determining biomass and grain yield. Leaf angle (plano-or erecto-phile) in canopy contributes to variation in interception.

Total number of bright sun-shine hours received during July-September in the wet season in different regions of India are shown in Fig. 4. Traditional rice growing areas receive only about 400 hours of sun shine in eastern India and 400-600 hours in southern India. In the non-traditional rice areas of north-western India, the sunshine hours during the same period are 600-700 hours.

Solar radiation requirements of a rice crop differed between vegetative and reproductive stages (Yoshida and Parao, 1976). While shading during the vegetative phase had a small effect on yield and yield components, shading during reproductive phase had a large effect on spikelet number and harvest index. Relationship between yield and solar radiation at different growth stages (Fig. 5) showed high association between 100 and 450 cal $cm^{-2}day^{-1}$ during reproductive phase and between 150 and 400 cal $cm^{-2}day^{-1}$ during

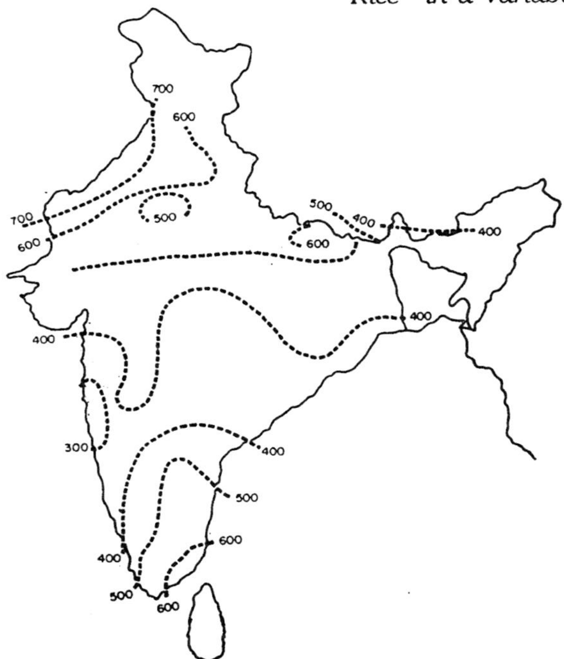

Fig. 4: Total number of bright sunshine hours during July-September in different regions of India. (Murty and Sahu, 1987)

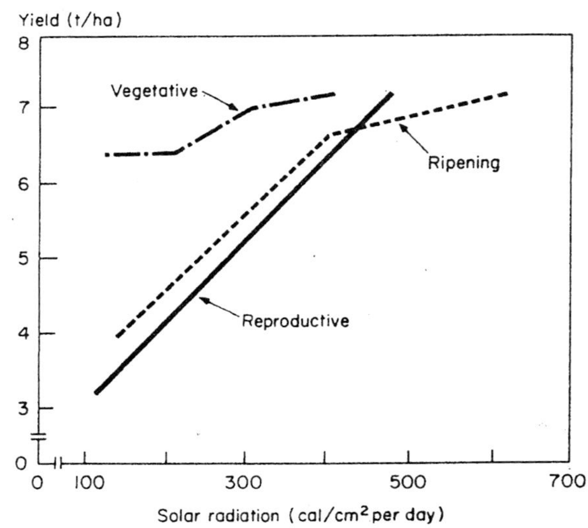

Fig. 5: Effect of solar radiation at different growth stages on grain yield of rice (Yoshida, 1981).

ripening phase. Only a weak relationship was observed between yield and 200 cal $cm^{-2}day^{-1}$ during vegetative phase.

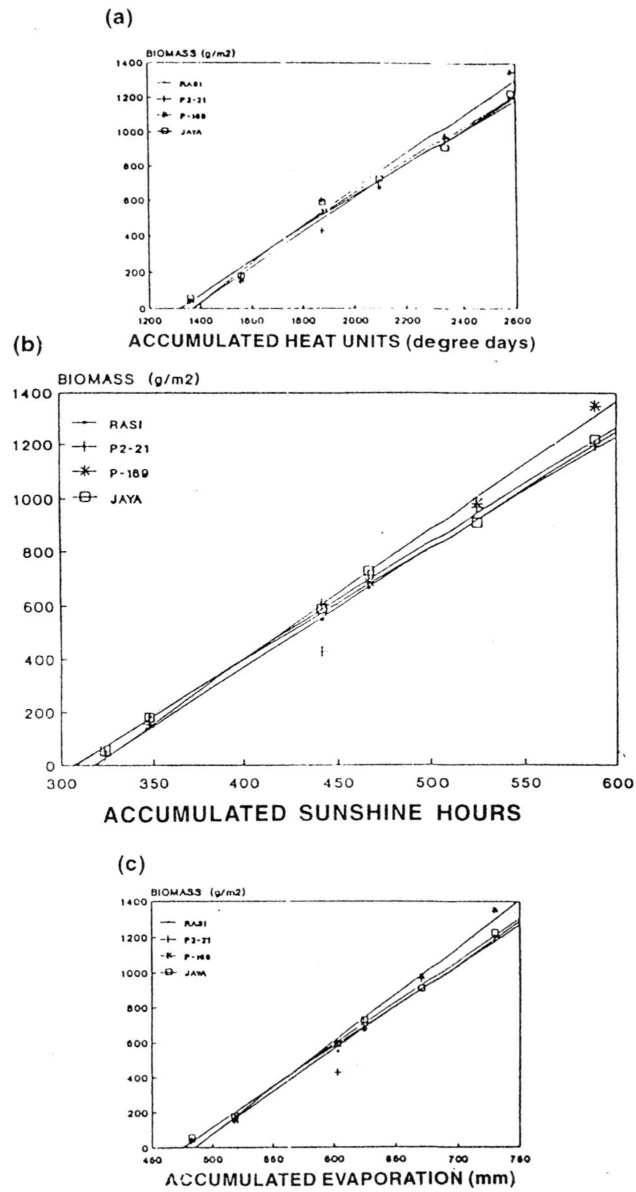

Fig. 6: Relationship between shoot biomass of rice and energy summation indices.

Grain yield was associated with total solar radiation available during reproductive, flowering and grain ripening phases (De Datta and Zarate, 1970; Evans and De Datta, 1979; Seshu and Cady, 1984). A positive relationship between 300-600 cal $cm^{-2}day^{-1}$ radiation during post flowering and 50 to 180 filled grains per panicle was found by Oldeman *et al.* (1987). Vamadevan and Murty (1976) concluded from their studies that cumulative solar radiation of 14,000 g-cal cm^{-2} or 200 hours of sunshine during 30 days preceding to harvest is optimum for obtaining miximum grain yield.

Shading resulted in rice leaves having thinner mesophyll, smaller individual cell volume and smaller cell number per unit leaf area (Chonan, 1967). Low light intensity from panicle initiation to flowering or from flowering to harvest reduced grain yield (Venkateswarlu, 1977). The period from panicle initiation to 10 days prior to maturity was found to be a critical light requiring period (Stansel, 1975). Murty and Murty (1982) showed that low ligh intensity from 10 days before anthesis to 20 days after anthesis induced high spikelet sterility leading to poor grain yields. Low light stess reduced grain number per panicle in short duration rice varieties, increased spikelet sterility in medium duration varieties and decreased panicle number in long duration varieties (Murty and Sahu, 1987). Spikelet sterility under low light at flowering was associated with decreases in carbohydrate content, protein synthesis, and cytokinin accumulation and increases in gibberellins and soluble N in the panicle. Number of empty spikelets increased with shading or lower intensity of solar radiation (Wada *et al.* 1973).

The influence of temperature and solar radiation on crop growth and yield cannot be fully separated, These two weather variables are inter-related and the responses of the crop are integrated. Shoot biomass accumulation was in fact closely related to three energy summation indices namely accumulated heat units (degree days), accumulated sunshine hours and accumulated evaporation (Fig. 6). The correlation coefficients were highly significant ($R^2 = 0.90$ to 0.99). At optimum combination of spacing and N levels, number of panicles/m^2 and spikelet number/m^2 were strongly influenced by levels of solar radiation and temperature during vegetative and reproductive growth (Yoshida, 1981).

8.2.3. Rainfall

Rainfall amount and distribution is the most critical weather component in rainfed rice ecologies (upland, lowland and flood-prone). Its effect on irrigated ecology is indirect through availability of water in irrigation sources (tank, canal, well, etc) and its effect on evapotranspiration of the crop through changes in temperature, solar radiation, vapour pressure deficit and wind speed.

8.2.4. Drought

Availability of about 200-300 mm of water per month is considered minimum to produce a good crop of rainfed rice. Drought is invariably a limiting factor affecting crop growth at any stage under upland conditions; drought may also occur in the early part of cropping season in the rainfed lowland and even in the flood-prone ecology, affecting seedling establishment and early vegetative growth.

The common symptoms of water deficit are leaf-rolling, leaf-scorching, impaired tillering, stunting, delayed flowering, spikelet sterility and poor grain filling. Water deficit during vegetative stage reduces plant light, tiller number and leaf area but the crop can recover without much loss in yield if water is available before flowering. Rice is most sensitive to water deficit from reduction division stage (10-11 days before flowering to one week after flowering). Drought during this time causes high spikelet sterility.

8.2.5. Submergence

Excess rainfall results in submergence of crop in rainfed lowland and flood-prone ecologies. The water level in fields, duration of submergence, and time of occurrence in relation to crop growth phase vary from season to season. The plants are either completely or partially submerged. Partial submergence is common problem in most of the rainfed lowland areas. Yield reduction is due to impaired tillering and decreased leaf area for photosynthesis. The decrease in yield depends on extent and duration of submergence.

8.2.6. Photoperiod

Length of photoperiod is not really a critical weather component because there is practically no fluctuation in it at a location unlike daily temperature, solar radiation and rainfall. The trend in the evolution and domestication processes of the rice has been toward a gradual loss of the strength of photoperiod and an increase in the basic vegetative phase (Chang, 1976). The modern high yielding varieties of early and medium duration are photoinsensitive. The long duration cultivars are usually photosensitive. Rice is a short-day plant. In flood prone ecology, photosensitive cultivars are grown. The proportion of sensitive varieties increases as the latitude decreases (Change and Oka, 1976). The critical day length also becomes shorter in the direction of south from temperate region.

8.2.7. Vapour Pressure Deficit

High vapour pressure deficit is associated with high temperature. This weather factor may aggravate water stress during drought under upland conditions. In dry season, vapour pressure deficit is high during grain filling phase. Even though the crop is irrigated, combination of high temperature and vapour pressure deficit may reduce stomatal conductance during mid-day, resulting in decrease in photosynthesis.

8.2.8. Wind

High wind velocity may cause lodging particularly in uplands. Physiologically, fluttering of leaves makes very little difference in the boundary layer resistance. Although photosynthesis may be expected to increase due to decrease in the boundary layer resistance, the resistance may typically decrease by 10 % and the consequent increase in photosynthesis rate is unlikely to exceed 1 % (Dr. J.L. Monteith's comments in Vergara 1976). Dry winds may cause desiccation of rice leaves in uplands. The mechanical damage of leaves by wind was more severe under upland than under lowland conditions (Alluri *et al.* 1973). Upland cultivars had thicker leaves or higher specific leaf weight and differed markedly in their resistance to wind damage. Plants grown at low levels of N were found to have high resistance to wind damage in terms of percentage of ripened grains and the weight and thickness of grains. The resistance was ascribed to accumulation of carbohydrates in plants under low N levels (Matsuzaki *et al.* 1972).

The review of the results of the studies on yield responses to weather components bring out that, among all weather components, the most important ones are temperature and solar radiation, if water is not limiting. When rice is subjected to stress due to low or high temperature, low solar radiation or drought (water deficit), spikelet filling percentage is the most sensitive yield component.

8.3. Responses to Climate Change

8.3.1. Elevated CO_2

The atmospheric CO_2 concentration has increased at a rate of 2.4 % per year during the last 50 years (Goudriaan, 1995). CO_2 influences plant growth directly because it is an input in photosynthesis. Numerous experimental observations indicate that increasing CO_2 will significantly influence photosynthesis, growth and yield of crops (Kimball, 1983; Baker *et al.* 1997b). Elevated atmospheric CO_2 increased the growth rate and yield of rice through enhanced photosynthesis and the degree of enhancement was influenced by light intensity, temperature and nutrition (Imai 1995). The increase in photosynthesis was 36 % more under 660 $\mu mol\ mol^{-1}$ than under 330 umol mol^{-1} CO_2 (Baker and Allen, 1993). Rice leaf photosynthetic rate increased with increasing CO_2 concentration upto 500 $\mu mol/mol^{-1}$ above which it tended to pleateau (Baker *et al.* 1990). CO_2 assimilation rates as a function of intercellular CO_2 in rice plants grown at 330 and 660 $\mu mol\ mol^{-1}$ was attributed to acclimation of Rubisco.

Enhanced CO_2 environment probably resulted in suppression of Rubisco activity, excessive accumulation of starch in chloroplasts, limited RUBP regeneration due to insufficient inorganic phosphate influx or limited sink size of plants relative to enhanced photosynthesis (Imai, 1995). Rubisco activity in rice, expressed on a leaf area basis, decreased by 66 % across 160-900 μmol mol^{-1} CO_2 in long-term CO_2 treatments (Rowland-Bamford *et al.* 1991). The decline in Rubisco activity was attributed by Baker and Allen (1994) to decrease in amount of Rubisco relative to other soluble proteins (Table 2).

Table 2: Effect of enhanced CO_2 concentration on leaf nitrogen and Rubisco protein (%) in 75-day old rice plants.

CO_2 concentration (umol mol^{-1}	Leaf N (mmol/m^2 of leaf area)	Rubisco protein (% of leaf soluble proteins)
160	95	62
250	90	59
330	81	54
500	62	49
660	78	43
900	64	42

Baker and Allen (1994) and Rowland-Bamford *et al.* (1991)

However, Ziska and Termura (1995) reported that enhancement of photosynthesis by about 50 % persisted in response to elevated CO_2 (600 μmol mol^{-1}). The response of assimilation to internal CO_2 concentration indicated an increased sensitivity of acclimation of Rubisco with an enhanced CO_2 environment. Photochemical efficiency (variable to maximal flourescence, Fv/Fm) was not significantly affected but dark respiration of leaves was reduced in plants grown under elevated CO_2 compared with plants grown under ambient CO_2 concentration. Suppressed dark respiration may increase rice productivity unless low respiration rate affects energetic processes (Imai, 1995). At elevated CO_2, respiration rate may be decreased by relative increase of the C-N ratio when maintenance respiration becomes a larger fraction (Amthor, 1991).

Elevated CO_2 significantly increased total plant biomass, mostly contributed by increase in stem and root weight (Ziska and Teramura 1995). Leaf weight did not change and increase in root weight was more than that of shoot indicating more biomass allocation to roots under high CO_2 . Baker *et al* (1990) also reported increase in root shoot ratio with increase in CO_2 concentration. Number and length of crown roots increased with elevated CO_2 because of increased partitioning to roots and a large root system might prevent lodging at high CO_2 and improve nutrient absorption (Imai, 1995).

Rice plants grown at high CO_2 had more tillers than plants grown at ambient CO_2 as supply of carbohydrate increased but leaf number per stem tended to decrease (Imai, 1995). Leaves accumulated carbohydrate more than mineral nutrients at high CO_2. Tillering is important for total leaf area development as individual leaf size is not increased with elevated CO_2. Leaf area development was not accelerated as much as dry matter under high CO_2 (Baker *et al.* 1990). Accordingly, specific leaf weight increased often at high CO_2 as a result of thicker leaves or increased starch or non-structural carbohydrate content of leaves (Imai, 1988). The ratio of starch to sucrose concentration increased linearly with CO_2 concentration (Rowland-Bomford *et al.* 1990).

High CO_2 promotes earlier flowering. Growth duration of rice was shortened by 10-12 days over a CO_2 treatment range of 160-500 μmol mol^{-1} due to shortened vegetative phase of development including reduction in number of leaves on the main stem (Allen *et al.* 1995).

Grain yield of rice increases under elevated CO_2. Yoshida (1976) reported that pre-flowering enrichment increased grain yield by increasing spikelet number per m^2, and to a lesser extent, weight per grain. In post-flowering enrichment, the yield increase resulted from increased spikelet filling % increased weight per grain. In a dry season, pre-flowering enrichment increased grain yield by 30 % and post-flowering enrichment increased grain yield by 10 %. The effective growth stage for CO_2 enrichment before flowering was found to be from neck-node differentiation (33 to 24 days before flowering) to spikelet differentiation stage (24 to 11 days before flowering). CO_2 enrichment before flowering increased sugar and starch % in leaf sheath and culm. Baker *et al.* (1990) reported that while individual grain mass was unaffected, increase in number of panicle/plant was primarily responsible for yield increase. Grain number per panicle may increase because high CO_2 may reduce flower degeneration during panicle development (Imai; 1995). Ripening percentage increase is brought out by increased assimilate availability. Harvest index does not show any trend with increasing CO_2 concentraion.

Water use efficiency (ratio of photosynthesis to transpiration) increases at high CO_2 concentration mainly due to increased phothosythesis (Drake and Melee, 1997) and smaller decrease in transpiration (Allen *et al.* 1995). As high CO_2 decreases stomatal conductance, transpiration may decrease if leaf temperature does not increase (Imai and Okamoto-Sato, 1991). Baker *et al.* (1997 a, b) concluded that increase in

CO_2 may provide a modest reduction of 10% in water used and thus increased drought avoidance.

Ziska and Teramura (1995) observed intraspecific variation in reproductive potential in response to enhanced CO_2. In one variety (Fujisaka-5), tiller number increased but not panicle weight; in another variety (IR 36), panicle weight was greater but tiller number remained the same. These differences in yield components indicate that rice genotypes could be selected for optimal yield in elevated CO_2 environment. The optimal yield is linked to degree of photosynthetic enhancement over time, total biomass and biomass partitioning to yield components.

8.3.2. Elevated CO_2 and Temperature Interaction

Global warming due to rising CO_2 and other green house gases is predicted (IPCC, 1996). Both CO_2 and temperature influence crop growth and so understanding of crop responses to interaction of elevated CO_2 and temperature would be more meaningful. A study on interactive effect of CO_2 enrichment and temperature on carbohydrate accumulation and partitioning showed that optimum temperature for total non-structual carbohydrate accumulations was same (25°C) for ambient and elevated CO_2 leaves but CO_2 enriched plants could survive and maintain carbohydrate production rates at higher temperatures than non-enriched plants (Rowland-Bamford *et al.* 1996). Canopy photosynthesis responses to light intensity (200 to 1600 $\mu mol\ m^{-2}\ s^{-1}$) were similar for temperature from 25/18°C to 37/30°C at 330 or 600 $\mu mol\ m^{-2}$ CO_2 (Allen *et al.* 1995). Positive CO_2 X temperature interaction on leaf photosynthesis of rice plants was observed within limits by Imai and Okamota-Sato (1991) but Baker and Allen (1993) reported relatively insensitive interaction under dense population.

Rate of tiller appearance was greatest with increase in temperature (37/29°C) and CO_2 (Ingram *et al.* 1995). However, the high temperature (37/29°C) shortened tillering duration. Relatively lower temperature treatment (33/25°C) produced tillers for a longer duration. Tiller development increased as CO_2 increased from 350 to 425 $\mu mol\ mol^{-1}$. Therefore, final tiller number depends on an optimum temperature under elevated CO_2 concentration. High night temperature and high CO_2 may lead to late tillering (Imai *et al.* 1985).

Rice ontogenesis was shortened both by elevated CO_2 treatment and by moderately high temperature (Allen *et al.* 1995). Earlier flowering induced by elevated CO_2 concentration might be further accelerated by rising temperature. Pronounced earliness may result in smaller plants and decrease in yield (Imai *et al.* 1985). Temperature ranging from 21 to 37°C greatly influenced developmental rate and total growth duration while the effects of CO_2 enrichment (330 and 660 μmol mol^{-1}) on these were relatively small (Baker *et al.* 1992).

Panicle emergence duration in *indica* varieties increased with increase in temperature at ambient CO_2 but decreased at higher temperature at elevated CO_2. Direct beneficial effects of CO_2 on growth may be partly negated by shortening of time to flowering (Ingram *et al.* 1995).

Shoot and root dry weight at flowering decreased as temperature increased from 29/21°C to 41/33°C (Ingram *et al.* 1995). This response contrasts with response of increase in shoot and root dry matter under elevated CO_2 levels.

Rice is most susceptible to damage by high temperature injury during flowering when temperatures of 38°C and above impair pollen shedding and fertilization (Mackill *et al.* 1982). Increasing temperature above 28/21°C resulted in decreased grain yield due to decline in filled spikelet % (Allen *et al.* 1995). Grain yield was higher at a mean temperature of 26°C and it declined by about 10 % for each 1°C rise in temperature above 26°C. The decrease in spikelet fertility (Fig. 7) and the decline of yield at higher temperature were not ameliorated under elevated CO_2 (Baker and Allen, 1993; Horie, 1993). However, the simulated yield increase at an elevated CO_2 concentration was more at higher solar radiation energy (Fig. 8) indicating that adverse effect of high temperature on yield might be partially compensated by elevated CO_2 at higher light intensity.

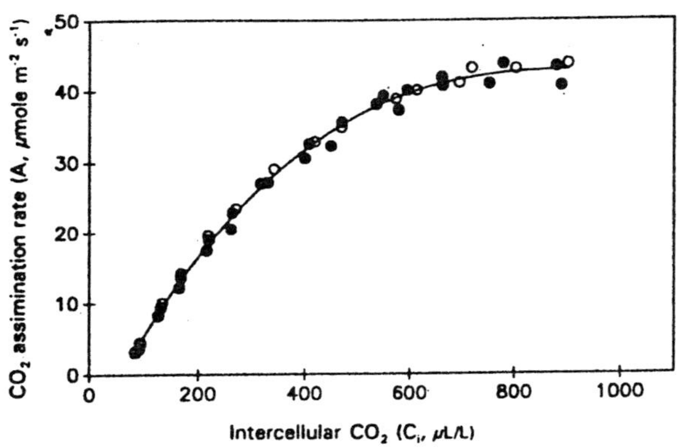

Fig. 7: Photosynthesis CO_2 assimilation rate (A) as a function of intercellular CO_2 (C) for single, attached, fully expanded leaves of rice plants grown at CO_2 concentrations of 330 μl/l (0) or 660 μl/l (O) and dry bulb air temperatures (day-night) of 35/25°C. Measurements were made at an irradiance of 1200-1300 μmol (photons) m^{-2} S^{-1} (Allen *et al.* 1995)

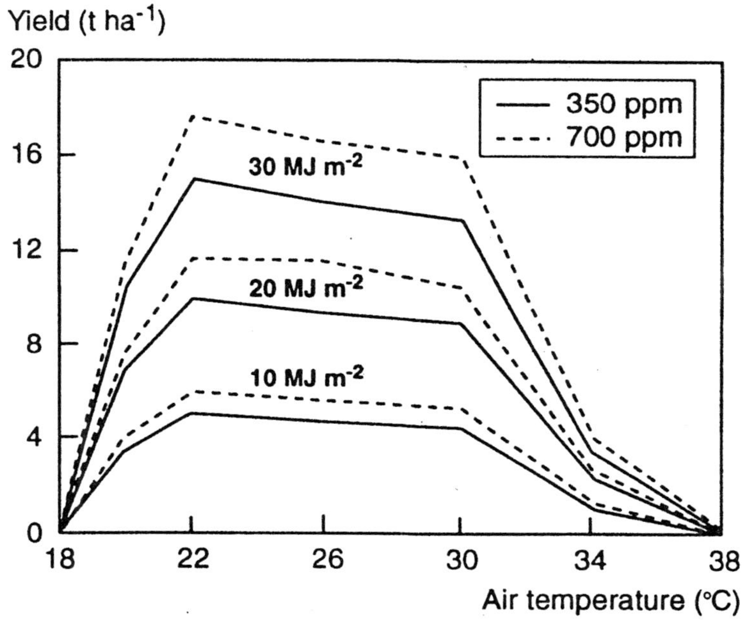

Fig. 8: Simulated yield responses of a rice variety to daily mean temperature, solar radiation, and CO_2 concentration, under constant environmental conditions. Day length was set at 12 h in each case. (Horie *et al.* 1995)

Rice cultivars vary in their tolerance to high temperature stress (37°C) during flowering and the ability of elevated CO_2 to mitigate high temperature effect on filled spikelet percentage was inversely related to apparent heat tolerance under ambient CO_2 (Ingram *et al.* 1995).

Water use efficiency increases with increasing CO_2 because of decline in water use and increase in photosynthesis but increasing temperature decreased water use efficiency because of increased water use and decreasing seed yield (Allen *et al.* 1995). Rising atmospheric CO_2 may benefit rice production by increasing photosynthesis, growth, and grain yield with reduction in water use. But, global warming in tropics may result in yield decreases with increase in water use.

Grain yield and quality of rice, partitioning, nutrient use efficiency and water use efficiency may be affected by elevated CO_2 x temperature interations (Conroy *et al.* 1994). Detailed field investigations will enchance our understanding on these important physiological aspects of rice production. Table 3 presents a summary of physiological, phenological, growth and yield responses of rice to elevated temperature and CO_2 level.

Table 3. Physiological, phenological, growth and yield responses of rice to increasing temperature or CO_2 concentration

Parameter	Increasing temperature	Increasing CO_2
1. Photosynthesis	No effect (25–25° C) Decrease (above 35° C)	Increase (levels off above 500 µmol mol^{-1})
2. Respiration	Increase (till 32° C) Decrease (above 32° C)	Decrease (levels off above 500 CO_2 levels)
3. Transpiration	Increase	Decrease (small effect)
4. Stomatal conductance	Complex response but decrease at supraoptimal	Decrease
5. Days to flowering	Decrease	Decrease
6. Grain filling duration	Decrease	No effect
7. Growth rate	Increase (22–31° C) Decrease (above 32° C)	Increase
8. Phyllochron interval (days/leaf)	Decrease	No significant effect
9. Tiller number	Decrease (above 37° C)	Increase
10. Leaf area index	No trend	No trend
11. Dry matter accumulation	Decrease (above 30° C)	Increase
12. Partitioning to root	Decrease (above 35° C)	Increase
13. Grain yield	Decrease (above 26–28° C)	Increase (levels off above 500 µmol mol^{-1})
14. Yield components		
i. No. of panicles/m^2	Decrease (above 35° C)	Increase
ii. No. of spikelets/panicle	Decrease	Increase
iii. Spikelet filling %	Decrease (above 32° C)	Increase
iv. Weight/grain	Decrease	Increase or no effect
15. Harvest index	Decrease	No effect
16. Water use efficiency	Decrease	Increase

8.3.3. Effects of UV-B Radiation on Rice

In the tropics of Asia, the daily level of UV-B radiaion can be as high as 11kJ/m^2 (Bachelet *et al.* 1991). Further increase in UV-B radiation resulting from stratospheric ozone depletion may affect rice production. Increased UV-B radiation adversely affected growth and productivity of plants (Teramura, 1983). UV-B radiation reduced activities of Rubisco and ATP-synthase, and inhibited reactions in photosystem II reaction centre, thus depressing photosynthesis and affected the levels of pigments, nucleic acids and polypeptide sub-units of Rubisco (Jordan *et al.* 1992; Strid and Porra, 1992). High UV radiation inhibited photosynthesis rate in rice (Kumagai and Sato, 1992).

Decrease in chlorophyll content was observed only under enhanced UV-B and low visible light (Tevine and Teramura 1989). In fact, chlorophyll content on leaf area basis generally tended to increase (Dai *et al.* 1992; Nouchi and Kobayashi 1995). The increase in chlorophyll content on leaf area basis was due to increased leaf thickness under enhanced UV-B radiation. The UV absorbing compounds, including flavanoids tended to increase with increase in UV-B radiation (Dai *et al.* 1992; Teramura *et al.* 1991).

Reduction in dry matter in rice under UV-B was attributed to decrease in stomatal density and opening (Dai *et al.* 1995b). UV-B reduced stomatal density and opening in green-house grown cultivars. A direct effect of UV-B on stomata was inferred from greater reduction of stomata on the adaxial surface than on the abaxial surface under 3 or 4 weeks of UV-B exposure. Difference in plant dry matter between UV-B treated and control plants was significantly correlated with the reduction in stomatal opening ($r = 0.834^{**}$) and density ($r = 0.717^{*}$) on adaxial surface under UV-B treatment, indicating that reduction in dry matter of rice plants under UV-B resulted from decrease in stomatal density and opening.

UV-B reduced dry matter accumulation in rice grown under controlled condition (Dai *et al.* 1994; Teramura *et al.* 1991). Significant decrease in total biomass was recorded with increased UV-B radiation in about one third of 16 rice cultivars from 7 different geographical regions, grown in a green house for 12 weeks under conditions simulating 20% ozone depletion (Teramura *et al.* 1991). However, Nouchi and Kobayashi (1995) reported that rice is likely to be less sensitive to UV-B radiation than other crops. Under field conditions, they did not find any effect of UV-B radiation for about 8 weeks on growth of 17 cultivars from 8 different geographical regions of the World. Days to flowering, plant height, tiller number, panicle dry matter and total dry matter were not influenced by enhanced UV-B (100 % increase over ambient solar UV-B control). The daily UV-B irradiance in ambient sunlight ranged from 0.3 to 8.3 kJ/m^2

Results of field studies conducted in four consecutive seasons in 1992 and 1993 at IRRI showed that enhanced UV-B radiation (13 kJ/m^2, about 20% ozone depletion) had no effect on rice growth, yield, photosynthesis rate, stomatal conductance, transpiration rate and internal CO_2 concentration (Table 4) although it signficantly reduced many components of growth and morphology in green house (Dai *et al.* 1995a). Visible solar radiation is greater under field conditions. Plants grown under high PAR were less sensitive to UV-B (Warner and Caldwell, 1983). Visible irradiance aids in photorepair processes in plant tissues. Green house-grown plants had lower levels of UV absorbing pigments and much thinner cuticles compared with field-grown plants (Caldwell *et al.* 1983). For these reasons, reliable estimates of the effects of predicted UV-B increase can be obtained in field studies only (Nouchi and Kobayashi, 1995).

Table 4. Physiological and yield responses of rice to enhanced daily UV-B radiaion (13 kJ/m^2) in field.

Physiological/yield parameter	Control	Treated
1. Photosynthesis (μmol CO_2 m^{-2} s^{-1})	21.9	22.3
2. Stomatal conductance (mol CO_2 m^2 s^{-1})	1.48	1.55
3. Transpiration rate (mmol H_2O m^{-2} s^{-1})	16.2	17.2
4. Panicles/m^2	437	440
5. Filled spikelet (%)	77	76
6. 1000 gran wt. (g)	23.0	22.9
7. Yield (t/ha)	6.14	6.07

Physiological data are means of 54 observations at 5 different growith periods of IR 74 in 3 experiments in 1992 and 1993 seasons. Yield and its components are means of four consecutive experiments in 1992 and 93 seasons for IR 64, IR 72 and IR 74 (adapted from Dai *et al.* 1995b).

Rice cultivars differed in sensitivity to UV-B. Dai *et al.* (1994) studied response of 188 rice cultivars from various rice growing regions and ecosystems and concluded that cultivars originally from regions with higher ambient UV-B are not necessarily more tolerant to enhanced UV-B radiations. Kumagai and Sato (1995) also concluded from their studies with different ecotypes of *indica*, tropical *japonica*, and temperate *japonica* rices that resistance to UV radiation was not related to the geographical origins of rice cultivars. Sato and Kumagai (1993) examined 198 rice cultivars of five Asian rice ecotypes for differences in resistance to injurious effects of UV radiation. They found that varieties belonging to the same ecotype showed different sensitivities. Investigations of the inheritance to resistance to injuries caused by UV radiation are important for identifying genetic resources and development of cultivars resistant to UV (Kumagai and Sato, 1995).

8.3.4. Modeling the Impact of Climate Change on Rice Production

General circulation models (GCM), describing the dynamic processes in the earth's atmosphere have been used to perdict potential climate change scenarios (Grotch, 1988). According to GCM predictions, a doubling of the current CO_2 would bring about changes in precipitation patterns. The effects of such climatic changes on food production would have serious consequences in any society. As rice is a major crop in India, the effect of climate change on its production needs to be known for evolving any anticipatory strategies to avert possible disaster and to maintain and/or to improve the total production of rice.

Three GCM models, i.e., 1. Geophysical Fluid Dynamics Laboratory (GFDL), 2. Goddard Institute for Space Studies (GISS) and 3. United Kingdom meteorological Office (UKMO), predict that the current temperature range of 20 – 30°C over India would change to 25 – 35°C under twice the present CO_2 concentration, (Bachelet *et al.* 1995). Both GFDL and UKMO models predict increase in rainfall over nothern India. A reduction of the extent of the rice growing areas is a possibility if spikelet sterility would increase under predicted higher temperatures and if water reserves could not meet the increased evapotranspiration. However, Bachelet and Kropff (1995) predict significant increase in the irrigated ecosystem in eastern India based on a modelling study on the impact of climatic change on agroclimatic zones in Asia.

Rice potential production simulation models, ORYZA 1 and SIMRIW (Simulation model for rice-weather relation), were used to evaluate different climate change scenarios in which temperature and CO_2 were varied in fixed increments above the current temperatures both independently and in combination, as well as scenarios predicted

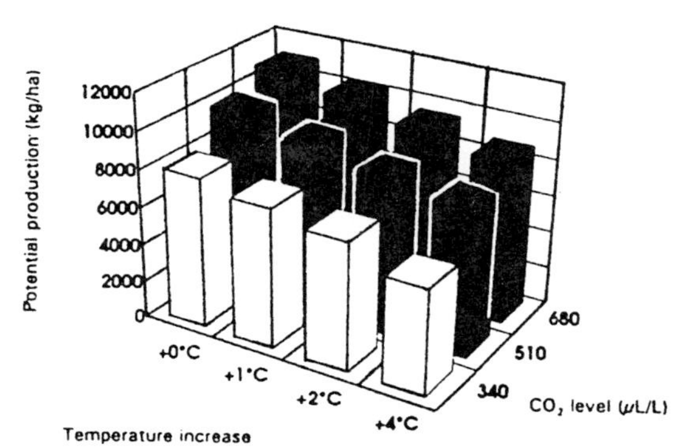

Fig. 9: The effects of increasing CO_2 level and temperature on potential yield of rice. (Mathews *et al.* 1995)

for a doubled CO_2 climate by the three GCM models (Mathews *et al.* 1995). The two crop simulation models predicted that at different CO_2 levels, an increase in temperature would cause decline in yields, and at all temperature increments, an increase in CO_2 would increase yields (Fig. 9). The average across both crop models and three GCM scenario indicated a - 3.4 % decline in overall rice production in Asia. But parts of India would be benefited from the climate change predicted by the three GCMs. The response of spikelet sterility to temperature has emerged as a major factor influencing the predictions of the climate change scenarios. Varietal adaptation to high temperatures and management practices such as advancing planting dates in the higher latitudes seem to be mitigation options.

The effects of fixed increments of temperature and CO_2 on potential yields were studied by Mohandass *et al.* (1995), using ORYZA 1 model, for prediction of potential rice production in India under future climate. When the temperature was not varied, increasing CO_2 concentration increased the yield in both wet and dry seasons (Table 5).

Table 5. Yield changes (%) predicted for wet and dry seasons in India under fixed scenarios of increased CO_2 and temperature.

CO_2 level (μmol mol^{-1})	Wet Season				Dry season			
	OC	+ 1C	+ 2C	+ 4C	OC	+ 1C	+ 2C	+ 4C
340	O	-5	-9	-13	O	-17.0	-34	-64
510	29	23	18	12	29	7	-15	-53
680	45	38	33	26	45	20	-4	-47

(Mohandass *et al.* 1995)

On the other hand, increasing temperature decreased the yield four times more in dry than in wet season. A doubling of CO_2 level would compensate for the detrimental effect of a 4^oC increase in temperature in yield in wet season but not in dry season.

Annual rice production has been predicted to increase in India under the three GCM scenarios, mainly due to an increase in yield of wet season. Although the predicted higher temperature would decrease the yield in dry season, the climate change may not have a large effect on total rice production in the country because less than 10 % of the area is under rice cultivation in dry season, contributing about 15% to annual rice production. However, it would still have significant local effect in the regions of India growing rice in dry season.

Under predicted global greenhouse warming condition, day-night temperature difference may decrease because night temperature will increase more than day temperature in rice growing areas due to large specific heat of water (Ziska and Teramura 1995).

Genetic variability for tolerance to high temperature and response to elevated CO_2 suggest that germplasm improvement is possible for meeting the challenge of climate change by improving adaptation of crops to changed climate (Ingram *et al.* 1995). Present warmer regions in tropics will have a much greater probability of adverse effects of a warming environment with higher CO_2 on food supply than temperate regions (Rosenzweig and Parry, 1994). Climate change impact will be likely region specific, depending on a region's air temperature range and water availability and due to significant interaction between CO_2 and temperature on crop growth and yield.

Acknowledgement

The authors are grateful to Mrs. Sudharani and S. Prabhakar of DRR for their assistance.

8.4. References

ADACHI, K. and INOUYE, J. 1972. Effects of photoperiod, light intensity, and components of culture medium on flower initiation in highly thermo-sensitive paddy rice plants. *Proc. Crop Sci. Soc. Jpn.* **41**: 78-82.

ALLEN, L.H. Jr., BAKER, J.T., ALBRECHT, S.L., BOOTE, K.J., PAN, D. and VU, J.C.V. 1995. Carbon dioxide and temperature effects on rice. In: *Climate Change and Rice* (eds. S. Peng *et al.*), Springer-Verlag, Berlin. pp. 258-277.

ALLURI, K., VERGARA, B.S. and VISPERAS, R.M. 1973. Observations on damage to rice leaves by strong winds. *Sabrao Newsl.* **5**: 129-132.

AMTHOR, J.S. 1991. Respiration in a future, high-CO_2 world. *Plant Cell Environ.* **14**: 13-20.

BACHELET, D. and KROPFF, M.J. 1995. The impact of climatic change on agroclimatic zones in Asia. In: *Modeling the impact of climate change on rice production in Asia* (eds. R.B. Matthews *et al.*), CAB International. U.K. pp. 85-94.

BACHELET, D., BARNES, P.W., BROWN, D. and BROWN, M. 1991. Latitudinal and seasonal variation in calculated solar ultraviolet-B irradiance (280-320 nm) for rice growing regions of Asia. *Photochem. Photobiol.* **54**: 411-422.

BACHELET, D., KING, G. and CHANEY, J. 1995. Climate change scenarios. In: *Modeling the impact of climate change on rice production in Asia* (eds. R.B. Matthews *et al.*), CAB International, U.K. pp. 67-83.

BAKER, J.T. and ALLEN, L.H.Jr. 1993. Effects of CO_2 and temperature on rice: a summary of five growing seasons. *J. Agric. Meteorol.* **48**: 575-582.

BAKER, J.T. and ALLEN, L.H.Jr. 1994. Assessment of the impact of rising carbon dioxide and other potential climate changes on vegetation. *Environ Pollut.* **83**: 223-235.

BAKER, J.T., ALLEN, L.H.Jr. and BOOTE, K.J. 1990. Growth and yield responses of rice to subambient, ambient and superambient carbon dioxide concentrations. *J. Agric. Sci.* **115**: 313-320.

BAKER, J.T., ALLEN, L.H.Jr. and BOOTE, K.J. 1992. Response of rice to carbon dioxide and temperature. *Agric. For. Meteorol.* **60**: 153-66.

BAKER, J.T., ALLEN, L.H.Jr., BOOTE, K.J. and PICKERINGS, M.B. 1997a. Rice response to drought under carbon dioxide enrichment I. Growth and yield. *Global Change Biology.* **3**: 119-128.

BAKER, J.T., ALLEN, L.H.Jr., BOOTE, K.J. and PICKERINGS, M.B. 1997b. Rice response to drought under carbon dioxcide enrichment 2. Photosynthesis and evapotranspiration, *Global Change Biology.* **3**: 129-138.

CALDWELL, M.M., ROBBERECHT, R. and FLINT, S.D. 1983. Internal filters: prospects for UV-acclimation in higher plants. *Physiol. Plant.* **58**: 445-450.

CARNAHAN, H.L., ERICKSON, J.R. and MASTERBROEK, J.J. 1972. Tolerance of rice to cool temperature—U.S.A. Rice breeding, IRRI, Los Banos, Philippines, pp. 535-540.

CHAMURA, S. and HONMA, M. 1973. Injury of low temperature to transplanted young seedling of rice plant (in Japanese). *Niigata Agric. Sci.* **25**: 1-9.

CHANG, T.T. 1976. Rice-*Oryza sativa* and *O.glaberrima.* Evolution of crop plants (Ed. N.W. Simmonds), Longman, London. pp. 98-104.

CHANG, T.T. and OKA, H.I. 1976. Genetic variousness in the climatic adaptation of rice cultivars. In: *Climate and Rice.* IRRI, Los Banos, Philippines. pp. 87-111.

CHAUDHARY, T.N. and GHILDYAL, B.P. 1970. Influence of submerged soil temperature regimes on growth, yield and nutrient composition of rice plant. *Agron. J.* **62**: 281-285.

CHONAN, N. 1967. Studies on the photosynthetic tissues in the leaves of cereal crops. IV. Effect of shading on the mesophyll structure of the rice leaves. *Proc. Crop. Sci. Soc. Jpn.* **36**: 297-301.

CONROY, J.P., SENEWEERA, A.S., BASRA, A.S., ROGERS, G. and NISSEN-WOOLER, B. 1994. Influence of rising atmospheric CO_2 concentrations and temperature on growth, yield and grain quality of cereal crops. *J. Plant. Physiol.* **21**: 741-758.

DAI, Q., CORONEL, V., VERGARA, B.S., BARNES, P. and QUINTOS, A. 1992. Ultraviolet-B radiation effects on growth and physiology of four rice cultivars. *Crop Sci.* **32**: 1269-1274.

DAI, Q., PENG, S., CHAVEZ, A.Q. and VERGARA, B.S. 1994. Intraspecific responses of 188 rice cultivars to enhanced UV-B radiation. *Environ. Exp. Bot.* **34**: 433-442.

DAI, Q., PENG, S., CHAVEZ, A.Q. and VERGARA, B.S. 1995a. Effect of enhanced ultraviolet-B radiation on growth and production of rice under green house and field conditions. In: *Climate Change and Rice* (eds. S. Peng *et al.*), Springer-Verlag, Berlin. pp. 189-204.

DAI, Q., PENG, S., CHAVEZ, A.Q. and VERGARA, B.S. 1995b. Effects of UV-B radiation on stomatal density and opening in rice (*Oryza sativa* L.). *Ann. Bot.* **76**: 65-70.

DE DATTA, S.K. and ZARATE, P.M. 1970. Environmental conditions affecting the growth characteristics, nitrogen response, and grain yield of tropical rice. *Biometeorology*. **4**: 71-89.

DRAKE, B.G. and GORZALES-MELEE, M.A. 1997. More efficient plants: A consequence of rising atmospheric CO_2. *Ann. Rev. Plant. Physiol. & Plant. Mol. Biol.* **48**: 609-639.

EVANS, L.T. and DE DATTA, S.K. 1979. The relation between irradiance and grain yield of irrigated rice in the tropics as influenced by cultivar nitrogen fertilizer application and month of planting. *Field Crop Res.* **2**: 1-17.

GOUDRIAAN, J. 1995. Global carbon cycle. In: *Climate Change and Rice* (eds. S. Peng *et al.*), Springer-Verlag, Berlin. pp. 207-217.

GROTCH, S.L. 1988. Regional intercomparisons of General Circulation Model predictions and historical climate data. US Dept. of Energy Report TRO41. Carbon dioxide Research Division, Washington, DC. DOE/NBB-0084, pp. 291.

HORIE, T. 1993. Predicting the effects of climate variation and effect of CO_2 on rice yield in Japan. *J. Agr. Meteor.* (Tokyo). **48**: 567-574.

HORIE, T., NAKAGAWA, H., OHNISHI, M. and NAKNO, J. 1995. Rice production in Japan under current and future climates. In: *Modeling the impact of climate change on rice production in Asia* (eds. R.B. Matthews *et al.*), CAB International, UK. pp. 143-164.

IMAI, K. 1988. Carbon dioxide and crop production. *J. Crops Sci. Jpn.* **57**: 380-391.

IMAI, K. 1995. Physiological response of rice to carbon dioxide, temperature and nutrients. In: *Climate Change and Rice* (eds. S. Peng *et al.*), Springer-Verlag, Berlin. pp. 252-257.

IMAI, K. and OKAMOTO-SATO, M. 1991. Effects of temperature on CO_2 dependence of gas exchanges in C_3 and C_4 crop plants. *J. Crop Sci Jpn.* **60**: 139-145.

IMAI, K. COLMAN, D.F. and YANAGISAWA, T. 1985. Increase of atmospheric partial pressure of carbon dioxide and growth and yield of rice. *J. Crop Sci. Jpn.* **54**: 413-418.

INGRAM, K.T., MANALO, P.A., NAMUCO, O.S., PAMPLONA, R.R. and WEERAKOON, W.M. 1995. Interactive effects of elevated carbon dioxide and temperature on rice growth and development. In: *Climate Change and Rice.* (eds. Peng *et al.*) Springer-Verlag, Berlin. pp. 278-287.

IPCC (INTERGOVERNMENTAL PANEL ON CLIMATE CHANGE). 1996. Introduction: Climate Change (eds. J.T. Houghton *et al.*), the IPCC scientific assessment. Cambridge University Press, Cambridge.

JORDAN, B.R., HE, J., CHOW, W.S. and ANDERSON, J.M. 1992. Changes in mRNA levels and polypeptide subunits of ribulose 1, 5-bisphosphate carboxylase in response to supplementary ultraviolet-B radiation. *Plant Cell Environ.* **15**: 91-98.

KANEDA, C. and BEACHELL, H.M. 1974. Response of *indica-japonica* rice hybrids to low temperatures. *Sabrao J.* **6**: 17-32.

KIMBALL, B.A. 1983. Carbon dioxide and agricultural yield: an assemblage and analysis of 430 prior observations. *Agron. J.* **75**: 779-788.

KUMAGAI, T. and SATO, T. 1992. Inhibitory effects of increase in near UV radiation on the growth of Japanese rice cultivars in a phytotron and recovery by exposure to visible radiation. *Jpn. J. Breed.* **42**: 545-552.

KUMAGAI, T. and SATO, T. 1995. Genetic study of resistance to inhibitory effects of ultraviolet radiation in rice. In: *Climate Change and Rice* (eds. S. Peng *et al.*), Springer-Verlag, Berlin. pp. 180-188.

MACKILL, D.J., COFFMAN, W.R. and RUTGER, J.N. 1982. Pollen shedding and combining ability for high temperature tolerance in rice. *Crop Sci.* **22**: 730-733.

MATSUZAKI, A., MATSUSHIMA, S. and TOMITA. T. 1972. Analysis of yield-determining process and its application to yield prediction and culture improvement of lowland rice. CXI. Effects of nitrogen restriction treatment in the middle growth stage of rice plants on the resistance to some disease and that to wind damages (in Japanese, English Summary). *Proc. Crop Sci. Soc. Jpn.* **41**: 402-408.

MATTHEWS, R.B., HORIE, T., KROPFF, M.J., BACHELET, D., CENTENO, H.G., SHIN, J.C., MOHANDASS, S., SINGH, S., DEFENG, Z. and LEE, M.H. 1995. A regional evaluation of the effect of future climate change on rice production in Asia. In: *Modeling the impact of climate change on rice production in Asia* (eds. R.B. Matthews *et al.*), CAB International, UK. pp. 95-139.

MOHANDASS, S. KAREEM, A.A., RANGANATHAN, T.B. and JEYARAMAN, S. 1995. Rice production in India under current and future climates. In: *Modeling the impact of climate change on rice production in Asia* (eds. R.B. Matthews *et al.*), CAB International, U.K. pp. 165-181.

MURTY, P.S.S. and MURTY, K.S. 1982. Effect of reduced light at different growth stages on the spikelet sterility in rice. *Madras Agric. J.* **69**: 121-123.

MURTY. K.S. and SAHU, G. 1987. Impact of low-light stress on growth and yield of rice. Weather and Rice. IRRI, Philippines. pp. 93-101.

NISHIYAMA, I. 1976. Effects of temperature on the vegetative growth of rice plants. Climate and Rice, IRRI, Philippines. pp. 159-189.

NOUCHI, I. and KOBAYASHI, K. 1995. Effects of ultraviolet-B radiation on growth of rice plants in the field. In: *Climate Change and Rice* (eds. S. Peng *et al.*), Springer-Verlag, Berlin. pp. 169-179.

OLDEMAN, L.R., SESHU, D.V. and CADY, F.B. 1987. Response of rice to weather variables. In: *Weather and Rice.* IRRI, Philippines. pp. 5-39.

OSADA, A., V. SASIPRAPA, M., RAHONG, M., DHAMMANUVONG, S. and CHAKRABANDHU, H. 1973. Abnormal occurrence of empty grains of *indica* rice plants in the dry, hot season in Thailand. *Proc. Crop Sci. Soc. Jpn.* **42**: 103-109.

OWEN, P.C. 1972. Effects of night temperature on growth and development of IR 8 rice. *Exp. Agric.* **8**: 213-218.

ROSEHZWEIG, C. and PARRY, M.L. 1994. Potential impact of climate change on world food supply. *Nature*, **367**: 133-138.

ROWLAND-BAMFORD, A.J., ALLEN, L.H. JR., BAKER, J.T. and BOOTE, K.J. 1990. Carbon dioxide effects on carbohydrate status and partitioning in rice. *J. Exp. Bot.* **41**: 1601-1608.

ROWLAND-BAMFORD, A.J., ALLEN, L.H. Jr., BAKER, J.T. and BOWES, G. 1991. Acclimation of rice to changing atmospheric carbon dioxide concentration. *Plant Cell Environ.* **14**: 577-583.

ROWLAND-BAMFORD, A.J., ALLEN, L.H. JR., BAKER, J.T. and BOWES, G. 1996. Interactions of CO_2 enrichment and temperature on carbohydrate accumulation and partitioning in rice. *Environ. Exp. Bot.* **36**: 111-124.

SASAKI, K. and WADA, S. 1973. Varietal difference in cold tolerance at different stages of panicle development in rice. *Hoku-No* (Japan) **40**(8): 7-14.

SASAKI, K., MAEDA, H. and WADA, S. 1973. Effects of temperature and nitrogen on the sterility at the meiotic stage of rice plants. Hoku-No (Japan) **40** (9): 1-7.

SATAKE, T. and YOSHIDA, S. 1978. High temperature-induced sterility in *indica* rices at flowering. *J. Crop Sci. Jpn.* **47**: 6-17.

SATO, K. 1972. Growth responses of rice plants to environmental conditions. I. The effects of air temperature on the growth at vegetative stage (in Japanese, English summary). *Proc. Crop Sci. Soc. Jpn.* **41**: 388-393.

SATO, T. and KUMAGAI, T. 1993. Cultivar differences in resistances to inhibitory effects of UV radiation in rice. *Physiol. Plant.* **91**: 234-238.

SATO, K., INABA, K. and TOZAWA, M. 1973. High temperature injury of ripening in rice plant. 1. The effects of high temperature treatments at different stages of panicle development on the ripening. *Proc. Crop. Sci. Soc. Jpn.* **42**: 207-213.

SESHU, D.V. and CADY, F.B. 1984. Response of rice to solar radiation and temperature estimated from international yield trials. *Crop Sci.* **24**: 649-654.

SREENIVASAN, P.S. 1980. Meteorological aspects of rice production in India. *Agrometeorology of Rice.* IRRI, Philippines. pp. 19-31.

STANSEL, J.W. 1975. Effective utilization of sunlight. Six decades of rice research in Texas. Research Monograph No. 4, Texas Agric. Exp. Stn. pp. 45-50.

STRID, A. and PORRA, R.J. 1992. Alteration in pigment content in leaves of *Pisum sativum* after exposure to supplemental UV-B. *Plant Cell Physiol.* **33**: 1015-1023.

TANAKA, I. and MUNAKATA, K. 1974. The influence of seasons on the growth of leaves and roots (in Japanese). *Proc. Crop. Sci. Soc. Jpn.* **43** (extra issue 1): 59-60.

TERAMURA, A.H. 1983. Effects of ultraviolet-B radiation on the growth and yield of crop plants. *Physiol. Plant.* **58**: 415-427.

TERAMURA, A.H., ZISKA, L.H. and SZTEIN, A.E. 1991. Changes in growth and photosynthetic capacity of rice with increased UV-B radiation. *Physiol. Plant.* **83**: 373-380.

TEVINI, M. and TERAMURA, A.H. 1989. UV-B effects on terrestrial plants. Photochem. *Photobiol.* **50**: 479-487.

VAMADEVAN, V.K. and MURTY, K.S. 1976. Influence of meteorological elements on productivity. In: *Rice Production Manual*, I.C.A.R., New Delhi, India. pp. 56-63.

VENKATARAMAN, S. 1987. Agrometeorological aspects of growth, yield and water relations with special reference to rice. In: *Weather and Rice*. IRRI, Philippines. pp. 47-54.

VENKATESWARLU, B. 1977. Influence of low light intensity on growth and productivity of rice. *Plant Soil.* **47**: 712-719.

VERGARA, B.S. 1970. Plant growth and development. *Rice Production Manual.* UPCA and IRRI, Philippines. pp. 17-37.

VERGARA, B.S. 1976. Physiological and morphological adaptability of rice varieties to climate. In: *Climate and Rice*, IRRI, Philippines. pp. 67-86.

VERGARA, B.S., CHANG, T.T. and LILIS, R. 1972. The flowering response of the rice plant to photoperiod. IRRI *Tech. Bull.* **8**: 1-55 (rev.).

WADA, S., KUNIHIRO, Y. and HONMA, A. 1973. Effect of some environmental factors during heading stage on the fertility of grains in rice plant. Bull. Hokkaido Prefect. *Agric. Exp. Stn.* **28**: 45-51.

WARNER, C.W. and CALDWELL, M.M. 1983. Influence of photon flux density in the 400-700 nm waveband on inhibition of photosynthesis by ultraviolet-B (280-320 nm) irradiation in soybean leaves: separation of indirect and immediate effects. Photochem. *Photobiol.* **38**: 341-346.

YOSHIDA, S. 1973. Effects of temperature on growth of the rice plant (*Oryza sativa* L.) in a controlled environment. *Soil Sci. Plant Nutr.* **19**: 299-310.

YOSHIDA, S. 1976. Carbon dioxide and yield of rice. In: *Climate and Rice*, IRRI, Philippines. pp. 211-221.

YOSHIDA. S. 1981. Fundamentals of rice crop science, IRRI, Philippines.

YOSHIDA, S. and PARAO, F.T. 1976. Climate influence on yield and yield components of lowland rice in the tropics. *Climate and Rice.* IRRI, Philippines. pp. 471-494.

ZISKA, L.H. and TERAMURA, A.H. 1995. Intraspecific variation in growth and photosynthetic responses of rice with increased carbon dioxide. In: *Climate Change and Rice* (eds: S. Peng *et al.*), Springer-Verlag, Berlin. pp. 244-251.

Rice—in a Variable Climate, 1999, pp. 151-166
Eds. Y.P. Abrol and Sulochana Gadgil

9. Using Simulation Models to Analyse Rice-weather Relationships

The strong links between the vagaries of monsoon and Indian agricultural production are well known. Such direct effects of climatic factors on crop growth and development have been subject of investigation. Several regression models have been developed to predict the relationship with agricultural productivity and its components. These agro-meteorological models when dealing with multi-year time series usually include a technology trend factor, thus lumping everything other than climatic factors in one regressor. However, such analysis where the effects of some specific factors is determined without consideration of interactions and feedbacks from other controlling elements can often be misleading. In addition to the climatic factors, there is a large number of edaphic, hydrologic, biotic, agronomic and socio-economic factors that influence crop growth and productivity. The magnitude of response to climatic factors may, for example, vary in different locations/experiments/ years because of genotypic interactions with climatic variability, applications of external inputs such as irrigation and fertilizers, pests and disease incidence, and agronomic management. A typical example is provided by studies relating to agro-ecological zoning crucial for planning agricultural development. Such studies have been heavily based on the rainfall statistics. Major quantitative output of such an analysis is growing days. From a purely ecological view point this may be adequate but it does not indicate much about the agricultural potential of the region. In addition, the growing period does not necessarily have a correlation with crop productivity. Growing period, for example, is more than 210 days in Assam and West Bengal, but the yield of principal rainy season crop, rice is very low due to limited radiation and input use. By comparison, the growing period is only 90-150 days in Punjab, yet rice yields are much higher due to significantly higher radiation and input usage. It is all the more difficult to relate the growing period with the production of post-rainy season crops. Perhaps this growing period based agro-ecological zoning was appropriate when agriculture was monoculture and rainfall as the principal cause of agricultural variability such as in large tracts of semi-arid tropics (Aggarwal, 1993). With the increasing availability of irrigation resources, and other technological changes, there is now a need to evaluate the integrated effect of various climatic factors including radiation and temperature which have a profound effect on crop growth and production. Recent studies have also shown that the relative contribution of rainfall to Indian crop production is progressively decreasing (Rao and Sinha, 1991) due to increasing contribution of technological change (improved irrigation, fertiliser use, pest control and new varieties).

Dynamic crop growth simulation is a relatively recent technique that facilitates quantitative understanding of the effects of dynamic climatic and edaphic factors, and agronomic management factors on crop growth and productivity in different land evaluation units. These explanatory models are quantitative description of the mechanisms and processes that result in growth of the crop. The processes could be crop physiological, meteorological, physical and chemical processes. Such a modelling assumes that the rate of change of system can be closely approximated by considering the rate of processes to be constant during short time periods. This is based on state variable approach in which current states (for example, weights of plant parts, evapotranspiration, leaf area index) are updated after every short interval (usually one day) considering the previous state and the rate which is influenced by internal crop properties and environment. This is repeated till the crop is mature. Figure 1 described the flow of material and information in such a crop model. Daily net photosynthesis is first calculated depending upon radiation, temperature, leaf area index, photosynthesis rate and maintenance respiration of the crop. The net assimilates are partitioned to different plant parts depending upon the phenological stage. The dry matter accumulated in leaves and stems influence photosynthetic area as well as sink capacity (number of grains). The partitioning of carbohydrates to grains and ambient temperature decide grain filling rate and grain yield. Water stress at a given time is determined depending upon demand (potential transpiration) and actual availability of water in the root zone. Nitrogen stress is estimated in a similar fashion on following the crop demand and soil supply principles. Stresses affect crop growth and development depending upon intensity and crop development stage. Availability of water and nitrogen in the soil profile is influenced by a number of factors as shown in figure 1. The effects of other nutrient deficiencies, help us in analysing the effect of various climatic factors while considering the interactions with other factors affecting crop growth and yield. Such an analysis is normally not possible with conventional experimental methods. More details on these processes and their interrelationships can be found in Keulen and Wolf (1986) and Penning de Vries *et al.* (1989). Most crop simulation models can be run on a PC/AT and XT computer.

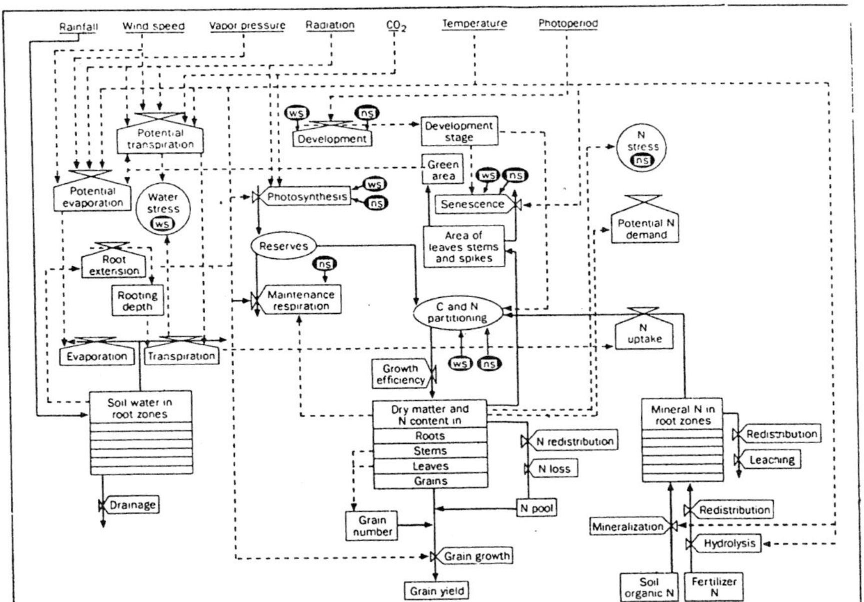

Fig. 1: A relational diagram explaining investigation and feedbacks among major physiological processes and their interaction with soil water and nitrogen variability in WTGROWS. Dashed lines indicate effect of climatic factors and information flow and solid line describe flow of materials Valves are rates, rectangles are quantities and circles are auxillary variables. To avoid complexity in representation, feedback loops of water and nitrogen stress is on crop physiological processes are shown as small circles only (Source: Aggarwal *et al.* 1994)

The models once developed are extensively tested for their performance in target environments and if found suitable are used for various applications. In order to correctly simulate crop behaviour models require precise data on weather (radiation, maximum and minimum temperatures, rainfall, vapour pressure deficit and wind speed), soil factors (available water, physical properties, depth), crop physiological properties (variety and few genoptypic constants related to the phenology, physiology and morphology of the crop/variety) and agronomic management (sowing date, plant population, amount and timing of irrigation and fertiliser applications). Since crop models integrate the effects of diferent factors on productivity they provide a unique opportunity to supplement results of field trials. The models have comparative advantage over conventional experiments and analysis since they can test numerous combinations (treatments) e.g. years (climatic variability)* variety * regions * water availability * sowing time * ...) which is not possible otherwise. None of the climatic variables have a 'fixed' value but vary with time. These computer based tools have the additional advantage of taking less time for analysing the effect of dynamic climatic variables and also taking care of the climatic variability.

A large number of crop models have been developed during the last three decades. Models are now available for all major crops such as wheat, rice, maize, sorghum, millet, chickpea, pigeonpea, groundnut, sunflower, sugarcane, potato and even for plantation and horticultural crops. Together with other tools of systems research such as Geographical Information Systems (GIS), databases and optimisation techniques, the models present a new opportunity for assessing potential production in a region and facilitate analysis of the sustainability options for agricultural development including planning of resource allocation. These approaches have been used in recent past for determining the production potential of a location knowing its resources, germplasm and the level of available technology, in matching agro-technology with the farmers resources and in analysing the precise reasons for yield gap, in estimating crop yield before the actual harvest, in studying short- and long-term consequences of agricultural practices on agriculture (Teng and Penning de Vries, 1992; Penning de Vries *et al.* 1993). In this paper, we briefly describe major rice models and some of their recent applications to analyse the effect of various factors, including climatic factors, on rice growth, development and productivity.

9.1. Rice Models

Several mechanistic rice models have become available during the last decade. All crop models primarily include the effect of temporal variation in climatic factors on growth and development. In addition, depending upon the objective, models include processes relating to crop growth, development, water and nitrogen uptake, and pest interactions. The following is a brief description of the three major rice models that have been well documented and validated.

9.1.1. Simriw

This is a simplified process based model for simulating potential growth and yield of irrigated rice (Horie *et al.*, 1992). It works on a daily time step and is driven by temperature and radiation. Absorbed radiation is converted to biomass using standard radiation use efficiency factors. Temperature affects phenological development, leaf area development and spikelet sterility. Phenology is also driven by phenology. Actual values of yield are obtained by multiplying the potential yield with a technology coefficient, the value of which has been determined for Japan and USA. The models performance has been found satisfactory for several key irrigated rice environments of Japan.

9.1.2. Ceres-Rice

This is a very detailed process model capable of handling the effect of weather, genotype, management, water and nitrogen on crop growth and productivity (*Singh et al.*, 1996). The model is designed to handle rice growth in various environments -lowland, upland, rainfed, irrigated, tropical and temperate. It requires daily values of radiation, temperature, rainfall and wind speed. Phenology is driven by temperature and daylength for panicle initiation. Radiation interception is simulated. Absorbed radiation is converted to biomass using standard radiation use efficiency factors. Dry matter is partitioned into various plant parts based on development stage dependent rules. Subroutines are included to simulate the uptake of water and nitrogen and its effect on crop growth processes. The performance of the model has been tested in various tropical environments in Philippines and India.

9.1.3. Oryza

This generic name represents a whole family of rice simulation models. ORYZA-1 is the basic physiological model which deals with crop growth, development and yield in potential production environment (Kropff et al., 1994). Like many earlier Wageningen based models, this model also simulated photosynthesis for the canopy integrated over the day, depending upon the sinusoidal variation in radiation and leaf area index dependent exponential attenuation of radiation. Temperature affects phenology, photosynthesis, maintenance respiration, leaf growth and spikelet sterility. The model includes th effect of carbon dioxide on canopy photosynthesis.

Several models have been evolved around this basic structure. ORYZA-N extends the capabilities of ORYZA-1 to handle crop N redistribution and its effects on crop growth in irrigated enviroments

(Drenth *et al.*, 1994). A summary model ORYZA-O simulates N uptake and is designed to develop N fertiliser management recommendations for irrigated rice (Berge *et al.*, 1994). Recently, Aggarwal *et al.* (1997) adapted ORYZA-a model to simulate soil N availability, uptake and effect of N stress on crop growth and yield.

Simulating rice response under water limiting condition brings into play several other factors which are not so obvious from figure 1. Rice is a crop which can be sown like any other upland crop in non-puddled soil (dry seeded rice). It can be also sown as pre-germinated seeds (wet seeded rice) or the seedlings transplanted (transplanted rice) in puddled soils. Moreover, it is grown under a diverse set of environmental conditions representing a large variation in soils, hydrological, climatic and edaphic conditions viz., irrigated, rainfed land, upland as well as flood prone deepwater ecosystems. Soil water relations, therefore, play an important role in simulating rice growth under such conditions.

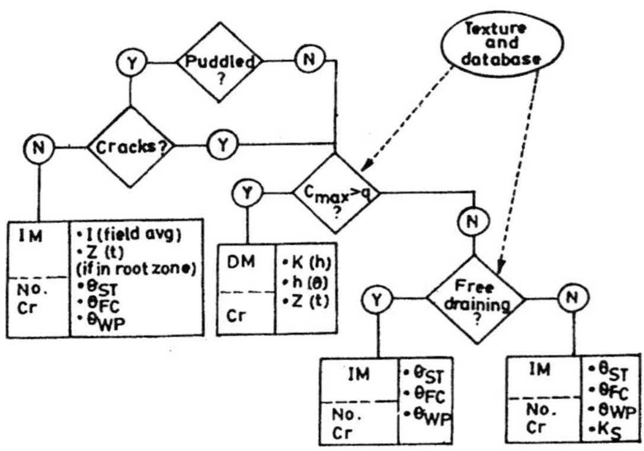

Fig. 2: General decision tree for selecting soil-water flow models. IM = Integral model; DM = Differential model; O_{ST} = water content at saturation; O_{FC} = water content at field capacity; O_{WP} = water content at wilting point; K_S = saturated hydraulic conductivity; Z(t) = groundwater table depth; I = infiltration; C_r = capillary flux to the root zone; C_{max} = Maximum capillary flux to the root zone; and, q = threshold value for capillary rise.

These conditions represent an even wider range of boundary conditions governing flow of water and it becomes essential to quantify the various components of soil water balance submodel, therefore, forms a very vital component in all models which simulate crop growth and development under water limiting conditions. As is evident from figure 1, the reduction in the potential transpiration under water limiting conditions is the key process which directly influences photosynthesis. Functions relating transpiration rate and other physiological processes to soil water potential have been derived for the various methods of rice establishment which make the model location independent (Singh *et al.*, 1995, 1996; Singh 1996) and which can be easily incorporated into a crop simulation model.

Soil water balanced models can be generally classified into two categories. One category consists of simple models which generally work on a balance sheet approach. They are called integral models and their data requirement is minimal. Many such models are available *e.g.* IR-RIMOD (Angus and Zandstra, 1980), PADDYWATER (Bolton and Zandstra, 1981). SAHEL (Penning de Vries *et al.*, 1989), PADDY (Wopereis, 1993) and ORYZA-W (Wopereis *et al.*, 1996) and depending upon the model, they can be used for puddled, non-puddled and freely draining soils. The complex models, also called as differential models, are based on the solution of the general flow equation. Moving from the integral to differential models increases considerably the data requirements. One typical example is SAWAH (Berge *et al.*, 1992). Measurements of some of the soil hydraulic properties required in differential models as input, like *in situ* hydraulic conductivity are time consuming, and costly. Use of simple integral models should be preferred over differential models until and unless the flow conditions are such that it is beyond the scope of the integral models.

These have been evaluated and flow conditions in which the simple models can be used without any loss in the output as well situations in which the differential models are unavoidable have been identified (Toung *et al.*, 1993) and shown in figure 2. Criteria for selection of a soil water balance model which include processes like preferential flow, soil cracking, internal drainage, significance of capillary rise effect of ponding depth etc. have also been discussed.

Unlike CERES-Rice, which can handle both water and N stress effects on crop growth at the same time, ORYZA family of models currently can not do so. The various models have been validated reasonably well in several tropical environments. The performance of ORYZA-1 has been found satisfactory even in temperate environments.

ORYZA models have been coupled with the quantified damage mechanisms to simulate the effect of pests on rice production. The BLIGHT model is a tool to analyse the effect of bacterial leaf blight and sheath blight (Ellings and Rubia, 1994). Similarly, another model-SBORER has been developed to mimic the damage caused to the rice plant by stem borers. Currently these models are being tested in the field (Ellings and Rubia, 1994).

9.2. Applications of Rice Modelling

9.2.1. Environmental Characterisation

Crop models together with GIS can greatly facilitate demarcation of homologous zones at mega-, macro-, meso- as well as micro level depending upon the availability of data and objectives. These tools have been used to determine potential and attainable yields for a given level of inputs for various crops. Potential yield of cultivars varies with season/year and location. Estimates of such yields for different varieties can establish a reference point for site quality and remove the confounding effects associated with large climatic variation. Matthews *et al.* (1995) used ORYZA 1 model to determine the potential yield in the main planting season of 4 tha^{-1} at Nanjing, China and the highest yields of 13 tha^{-1} at Kapurthala in India characterised by high solar radiation and low temperatures during the growing season. A good correlation was observed between main season potential grain yield and the photothermal quotient (PQ in MJ m^{-2} D^{-1} C^{-1}), the ratio of radiation and the mean temperature above the base temperature (8C) over the period from PI to flowering. The relationships were as follows :

Indica rice - Yield (kg ha^{-1}) = 526 + 9384 PQ (r^2 = 0.69)

Japonica rice - Yield (kg ha^{-1}) = –1563 + 9292 PQ (r^2 = 0.45)

Relatively low value of r^2 indicates that there are factors other than temperature and photoperiod as well responsible for determining yield differences across locations.

Thiyagarajan *et al.* (1993) conducted a simulation study to characterize the rice producing areas in various agro-climatic zones of Tamil Nadu in terms of potential yield of cultivars-ADT36 and CR1009. Their results showed that potential rice yields varied between 6.5 to 9.0 tha^{-1} in the Kuruvai season and between 5.5 and 9.0 tha^{-1} in the Samba season depending upon the location, year and transplanting date. The optimum transplanting time to produce more than 8 tha^{-1} yield differed for the various locations.

In tropical environments, grain yield is generally higher in the dry season compared to the wet season in well managed crops. In east Java, Indonesia, however, the reverse is true. Recent simulation studies have helped in explaning this phenomenon. The grain filling duration, critical for grain yield, was observed to be shorter in the dry season than in the wet season due to relatively higher temperatures in the dry season (Daradjat & Fagi 1991). This suggests the need to shift transplanting period in the dry season so as to avoid high temperatures during grain filling and yet exploiting higher radiation.

In rainfed environments, rice yields may be reduced due to drought. Wopereis et al. (1993) studied the potential of crop models to quantify risk based on soil hydrological and climatic data on a regional level for Philippines. Crop model ORYZA-W predicted potential rice yields to vary between 5.4 and 6.7 t ha^{-1} and the rainfed yields to vary from 0 to 6.7 t ha^{-1} thus quantifying the large effect of rainfall variablity on yield variability. The spatial distribution of risk was mapped using a GIS package. Angus (1989) used a rice model and historical weather data to determine the probabilities of yield above a specified threshold level for the Philippines varied little (3.5-4.0 t ha^{-1}) indicating rather uniform availability of water-stress free period on a long-term basis. However, the percentage of years in which two wet seeded rice crops could be simulated varied from less than 10 to 85 % among the simulated locations indicating considerable climatic variability.

9.2.2. Optimising Crop Management

Once potential yields have been quantified, these can be converted to attainable yields to determine magnitudes of yield gap. Crop growth modelling can be used to in matching agro-technology with the farmers resources and in analysing the precise reasons for yield gap. Recent studies have shown that simulation models can help fine tune the N fertiliser appication recommendations in irrigated rice (Berge *et al.*, 1996). Studies done in China, India, Indonesia and Philippines showed that simulation model recommendations helped in increasing the fertiliser use efficiency by 15-20 %. This is a very important result that needs to be tested on larger areas. Recent work done in China on a 10 ha extension farm plot has shown that following the N fertiliser recommendations generated using model ORYZA-0, rice grain yield was 9.4 % more than the one obtained by following the current national recommendations (Qinghua, S, personal communication). SImilar work is also being done in our country together with All India Project on Soil Test and Crop Response (Thiyagarajan, T.M.T., personal communication).

9.2.3. Pest and Disease Management

At a regional level, GIS, requisite environmental data coupled with epidemic simulation models further provide geographic delineation of disease and insect pest risk zones, such as those for rice blast. These zones can also help us in making strategic decisions on deployment strategies for varieties and to determine how long host plant resistance would be expected to last. Historical climate data from sites have been shown to be useful for characterising the conduciveness of a site to specific diseases. Attempts are also being made to integrate disease predictive systems with online weather and weather-interpolation systems.

Yield loss studies have conventionally quantified the relations between nitrogen application rate, disease severity, season and grain yield. This has resulted in a qualitative understanding of the host-pathogen-environment interactions and in disease management recommendations. However, validity of such recommendations is limited, as they are strongly influenced by disease onset, disease spreading rate, farm management practices, environmental conditions and their interactions. Physiologically based simulation models can be applied to understand the damage mechanisms and analyse their effects on crop growth and yield of rice. Attempts are being made in our country for such model application through generation of disease scenarios and effect of weather × nitrogen × disease interaction on bacterial blight.

9.2.4. Impact of Climate Change

Crop models -ORYZA1 and SIMRIW were recently used to estimate the impact of increased carbon dioxide and temperature on potential grain yield of rice in Asia (Matthews *et al*, 1995). It was assumed that the trends in potential yields would also be shown in actual yield. This assumption, may not be correct for environments where crop growth is limited by abiotic or biotic stresses such as in vast areas of India. Three GCMs were used to establish climate change scenarios. The crop models were driven by the weather data of 70 weather stations spread all over Asia. For India, data was limited to 9 weather stations.

The predicted change in rice yield varied from a 13 % to 7 % depending upon GCM used and rice model. The positive effects of increasing CO_2 decreased with the rise in temperature. The study on India showed that the annual rice production would increase in future as climate change occurs. This increase was due to increased productivity of main season crops where the fertilising effect of greater CO_2 compensated for the increased temperature associated decrease in yield in the second season (Mohandass *et al.*, 1995). The large decrease in the yield of second season crop could, however, be offset by the selection for more temperature tolerant varieties. The study has limited direct utility bacause it does not consider the effects of water availability and fertiliser use, two major determinants of rice productivity in India.

9.2.5. Yield Forecasting

Reasonably precise estimates of acreage and yield before the actual harvest are of immense value in policy planning. The relatively small cost and speed of assessment makes crop growth simulation models promising for areas where significant daily weather data are readily available. In this approach, the model is run using actual weather data during the cropping season for the geographical region of interest. Weather data for typical years are used to continue simulations until harvest. Horie et al. (1992) showed an example where crop models, regional weather databases and historical yield data is used to forecast rice yields for different regions of Japan. There is an urgent need to develop such models for Indian situation.

9.2.6. Optimal Sowing Dates for Hybrid Seed Production

A major constraint to utilizing hybrid seed technology is the need to buy fresh seeds every planting season and the high cost of seed production. This limited seed production is largely due to poor outcrossing and asynchronous flowering of male and female parents. Environmental factors such as temperature, humidity, and wind speed as the time of pollination and fertilization play a great role in regulating outcrossing percentage. To synchronise at several locations to determine the optimal planting dates. Crop simulation modelling is a more efficient tool to assist us in making decisions about the planting calender of parents in different environments. Such models are now available (Xu 1995). These models simulation and predict the development process of a variety per cent line use in hybrid seed production in a particular environment, as defined by its temperature and day length. Xu (1995) showed that simulation models predicted optimal planting dates for several parental lines that were very close to the planting dates determined by a large number of experiments. Ideally, dependable weather forecasts (temperature alone) should be available to make accurate predictions. Until good quality weather forecasts are available, reasonable guidance can also be obtained by using good quality, historical daily weather data for a few years for the target locations.

9.2.7 Increased Efficiency of Multi-environment Evaluation

Development and release of a variety is a complex process that may extend over a period of 10-15 years. Once the breeding lines have become homozygous, they are bulked and then tested in observational, replicated and multilocation yield trials. These multi-site trials are expensive and need several crop seasons/locations to understand genotype by environment (G × E) interactions.

In the state of Tamil Nadu in India, multilocation evalution of rice is done over a period of 10 years at 100 locations and a variety may be evaluated in as many as 288 experiments before it is released (Palanisamy *et al.*, 1993). Despite extensive testing done with new genotypes using the conventional approach, it is impossible to cover the whole range of environments. It may often happen that a specific environemntal challenge may not be available in natural environments in specific years to enable effective discrimination among breeding lines.

Since the systems approach integrates different components of agroecosystems, it can play a great role in increasing the efficiency of plant breeding process, in particular multi-environment testing of genotypes. Crop models together with GIS can facilitate biological characterization of the physical environment (geography, soil, climate, etc.), and thus define key environmental domains for which improved varieties are to be developed. Alternatively, the same methodology can be used to determine the adaptation domains of genotypes. A modelling approach can also provide estimates of yield probability in target environments based on our understanding of G × E interactions. Such studies can help in reducing the number of sites/season required for field evaluation and thus increase the efficiency of the whole process of variety development.

A methodology to asses the relative performance of breeding lines in a relatively short time has been proposed by Aggarwal *et al.* (1996). The necessary crop inputs can be measured at a few selected sites and the G × E interactions quantified over other sites. The methodology can be used to assess the performance of various genotypes across a much wider range of management options than would be possible by experimentation.

Palanisamy et al. (1993, 1995) assessed the performance (indicated by ranking) of pre-release long and short duration rice genotypes in several locations of Tamil Nadu. The average simulated ranking of genotypes across locations showed a reasonable agreement with the measured values. The first three lines identified by experimentation were also selected by simulation. Some variability in ranking is possible because the measured grain yields of different genotypes were rather close. For example, the difference in grain yield between the first and third genotype was only 0.28 t ha^{-1} and that between the first and last genotype as 0.78 t ha^{-1}. Such small differences between lines commonly observed in many well-conducted advanced multi-environment trials, are difficult to simulate with crop models largely because of inaccuracies in measurement of model imputs and because such difference are generally beyond the crop model's resolution.

9.3. Conclusions

A number of opportunities are now available for the use of crop simulation models for quantifying the effects of various factors including weather on agriculture. The key areas are biophysical characterization of agro-environments, evaluating impact of climate change, optimising crop, pest and disease management, and increasing the efficiency of multi-environment testing. We are now able to explain with simulation models the effects of environmental variability on crop growth and yield. The response of standard cultivars to environment can be predicted with confidence. In future, the systems approach, with its well-developed analytical framework, data bases, and powerful simulation models, will be handy to provide answers to many to the current agricultural issues in a relatively shorter time frame. To do this more effectively, greater efforts should be made to develop models that can integrate the effect of all important factors operating in the field environment, for instance. weather, edaphic conditions, management, incidence and effect of pests, and socio-economics.

Simulation modelling-based approaches generally require inputs of crop physiological data to characterize a genotype, soil characteristics, daily weather data, and management information. Many of these could be significantly uncertain due to our imperfect knowledge of their operation as well as due to random errors related with size and number of observations, and systematic errors related with bias in the experimental, measurement, observation, and calibration procedures. In addition, many input parameters exhibit spatial and temporal variability. This results in uncertainty in the outputs of deterministic crop models. There is an urgent need to document the characteristics of spatial and temporal aspects of climatic variability all over India. Establishment of an on-line agricultural and meteorological information system providing information not only of historical data but also on the current and future situation as well will go a long way in helping to understand the impact of climatic variability on agriculture and thus determining options for sustainable increase in agricultural production in future. This requires a close collaboration among meteorologists, agricultural scientists and modellers.

9.4. References

AGGARWAL, P.K. 1993. Agro-ecological zoning using crop growth simulation models; characterisation of wheat environments of India. p. 97-109, In; Penning de Vries, F.W.T. Teng, P.S. Melvelor (eds) *Systems Approaches for Agricultural Development.*, Kluwer Academic Publishers, Netherlands. pp. 542.

AGGARAWAL, P.K., KROPFF, M.J., CASSMAN, K.G. and BERGE, H.F.M. TEN. 1997. Simulating options for increasing irrigated rice yield potential. *Field Crops Research* **51**: 5-17.

AGGARWAL, P. K., KROPFF, M.J., TENG, P.S.and KHUSH, G. S. 1996. The challenge of integrating systems approach in plant breeding; opportunities, accomplishments and limitations. In: Kropff, M.J., Teng, P.S., Aggarwal, P.K., Bouman, B., Bouma, J. and van Laar, H.H. (eds.) p. 1-24 *Applications of Systems Approaches at the Field Level.* Kluwer Acad. Pub., Netherlands, pp. 465.

ANGUS, J.F. 1989. Simulation models of water balance and growth of rainfed rice crops grown in sequence. ACIAR Monograph, Australia.

ANGUS, J.F. and ZANDSTRA, H.G. 1980. Climatic factors and the modelling of rice growth and yield. Proc. WMO-IRRI Symposium on *Agrometeorology of the Rice Crop, International Rice Research Institute*, Los Banos, Philippines, pp. 189-199.

BERGE, H.F.M. ten, WOPEREIS, M.C.S., RIETHOVEN, J.J.M., THIYAGARAJAN, T.M. and SIVASAMY, R. 1994. ORYZA-0 model applied to optimise N use in rice. In: H.F.M. ten Berge, M.C.S. Wopereis, and J.C. Shin, (eds), *N Economy of Irrigated Rice: Field and Simulation Studies.* SARP Research Proceedings. IRRI, Los Banos, Philippines and AB-DLO, Wageningen, p. 235-253.

BERGE, H.F.M. TEN, AGGARWAL, P.K. and KROPFF, M.J. (eds) 1996. Systems modelling in rice research.Elsevier Publishers, Netherlands.

BERGE, H.F.M., TEN JANSEN, D.M., RAPPOLDT, K. and STOL, W. 1992. The soil water balance module SAWAH: Users guide and outline. CABO-PE Simulation Report Series, no. 22, CABO, Wageningen, The Netherlands, pp. 78.

BOLTON, F. R. and ZANDSTRA, H.G. 1981. A soil moisture based yield model of wetland rainfed rice. IRRI Research paper No. 62, IRRI,Los Banos, Philippines.

DARADJAT, A.A. and FAGI, A.M. 1991. Effect of seasonal variation in weather on crop potential of rice cultivar IR36 in the northe coastal region of west Java. 79-84. In: Penning de Vries, FWT, van Laar, H.H. and Kropff, M.J. (eds). *Simulation and systems analysis for rice production*. Pudoc, Wageningen, The Netherlands. pp. 369.

DRENTH, H., BERGE, H.F.M. TEN and Reithoven, J.J.M. (eds), 1994. ORYZA simulation modeules for potential and N limited rice production. SARP Research Proceedings. IRRI, Los Banos, Philippines and AB-DLO, Wageningen, pp. 223.

ELLINGS, A. and RUBIA, E.G. 1994. Models for the crop pathosystems. Rice-bacterial leaf blight, sheath blight and stem borer. SARP Research Proceedings. IRRI, Los Banos, Philippines and AB-DLO, Wageningen, pp. 102.

HORIE, T., YAJIMA, M. and NAKAGAWA, H. 1992. Yield forecasting. 211-236. In: Teng P.S., Penning de Vries FWT *Systems approaches for agricultural development*. Elsevier Applied Sciences, London. pp. 309.

Keulen, H. van and Wolf, J. (eds). 1986. Modelling of agricultural production: weather, soils and crops. Pudoc, Wageningen, Netherlands, pp. 479.

KROPFF, M.J. , VAN LAAR, H.H., MATTHEWS, R.B. (eds). 1994. ORYZA1: An ecophysiological model for irrigated rice production. SARP Research Proceedings, IRRI, Wageningen, pp. 110.

MATTHEWS, R.B., KRPFF, M.J. and BACHELET, D. 1995. *Modelling and impact of climate change on rice production in Asia*. CAB International, pp. 289.

MOHANDASS, S., KAREM, A.A. RANGANATHAN, T.B. and JEYRAMAN, S. 1995. Rice production in India under current and future climates. 165-181. In: Matthews, R.B., Kropff, M.J. and Bachelet, D. 1995. *Modelling and impact of climate change on rice production in Asia*. CAB International, pp. 189.

PALANISAMY, S., PENNING DE VRIES, F.W.T., MOHANDASS, S., THIYAGARAJAN, T.M., KAREEM, A.A. 1993. Simulation in pre-testing of rice genotypes in Tamil Nadu. Pages 63-75, in: Penning de Vries FWT, Teng P.S., *Metselaar K.* (eds). *Systems approaches for agricultural development*. Kluwer Academic Publishers, The Netherlands. pp. 542.

PALANISAMY, S., AGGARWAL, P.K., THIYAGARAJAN, T.T., RANGNATHAN, T.B.R. 1995. Simulating yield and ranking of rice genotypes in multi-location trials. In: Aggarawal P.K., Matthews, R.B., Kropff, M.J., van Laar, H.H.

(eds). *Applications of systems approach in plant breeding.* SARP Research Proceedings, AB-DLO, Wageningen and IRRI, Philippines. pp.

PENNING DE VRIES, F.W.T., JANSEN, D.M., TEN BERGE, H.F.M., BAKEMA, A.H., 1989. Simulation of Ecophysiological Processes in Several Annual Crops. Simulation Monograph. PUDOC, Wageningen, The Netherlands, pp. 271.

PENNINGS DE VRIES, F.W.T., TENG, P.S., METSELAAR, K. (Eds.). 1993. Systems Approaches for agricultural development. Kluwer Academic Publisher, Dordrecht and IRRI, Los Banos, pp. 542.

RAO, N.H. and SINHA, S.K. 1991. Sustainable development of water resources for agriculture. In: *Glimpses of science in India* (editor U.S. Srivatava). Malhotra Publishing Houe, New Delhi, p. 173-208.

Singh, A.K. 1996. Evaluation of models quantifying drought response in rice. Proc. 2nd International Crop Science Congress on 'Crop Productivity and Sustainability - Shaping the Future', New Delhi, Nov. 17-24, 1996, p. 50.

SINGH, A. K., TUONG, T.P., INGRAM, K.T., PERNITO, R. and SIOPONGCO, J. 1996. Drought response of dry seeded rice and its modelling. In '*Physiology of Stress Tolerance in Rice*', p. 186-197.

SINGH, A.K., TUONG, T.P. WOPEREIS, M.C.S., BOLING, A. and KROPFF, M.J. 1995. Quantifying lowland rice responses to soil-water deficit. p. 507-519. In: IRRI. Fragile lives in fragile ecosystems, IRRI, Los Banos, Philippines. pp. 976.

SINGH, U., RITCHIE, J.T. and GODWIN, D.C. 1993. A. User's Guide to CERES-Rice-V 2.10. *Simulation Manual* IFDC-SM-4 International Fertilizer Development Centre, Alabama, USA. 131.

SINGH. V.P., SINGH B.B., ZEIGLER, R.S. 1996 (eds.) Physiology of stress tolerance in rice NDAUT, IRRI, Los Banos, Philippines. pp. 239.

TENG, P.S., PENNING DE VRIES, F.W.T. 1992. Systems approaches for agricultural development. Elsevier Applied Sciences, London p. 309.

THIYAGARAJAN, T.M.T., JEYRAMAN, S., BUDHAR, M.N. and KAREEM, A.A. 1993. Rice yield potential in different agro-climatic zones of Tamil Nadu- a simulation study. p 181-190. In: Bouman, BAM, Laar, H.H and Zhaoqian, W. (eds). *Agro-ecology of rice based cropping systems*, SARP Research Proceedings, AB-DLO, Wageningen and IRRI, Philippines.

TUONG, T.P., WOPEREIS, M.C.S., SINGH, A.K., KROPFF, M.J. and TEN BERGE, H.F.M. 1993. Modeling lowland rice agrohydrology. Paper No. 932528 presented at the International Winter Meeting, Chicago, Dec. p. 14-17, 1993, p. 18.

WOPEREIS, M.C.S., 1993. Quantifying the impact of soil and climate variability on rainfed rice production. Ph.D. Thesis, Wageningen Agricultural University, pp. 188.

WOPEREIS, M.C., KROPFF, M.J., HUNT, E.D., SANIDED, W. and BOUMA, J. 1993. Case study on regional application of crop growth simulation models to predict rainfed rice yields: Tarlac province, Philippines. p. 27-46. In: Bouman, BAM, Laar, H.H. and Zhaoqian, W. (eds). *Agro-ecology of rice based cropping systems*. SARP Research Proceedings, AB-DLO, WAgeningen and IRRI, Philippines. pp. 201.

WOPEREIS, M.C.S., Bouman, B.A.M., Toung, T.P., Berge, H.F.M. and Kropff, M.J. 1996. ORYZA-W: Rice growth model for irrigated and rainfed environemnts. SARP Research Proceedings, AB-DLO, Wageningen and IRRI, Philippines. pp. 159.

XU, Q. 1995. A model for simulating hybrid rice seed production. p. 97-108. In: P.K. Aggarwal, R.B. Matthews, M.J. Kropff, H.H. van Laar (eds). *Applications of systems approach in plant breeding*. SARP Research Proceedings, AB-DLO, Wageningen and IRRI, Philippines. pp. 144.

Rice—in a Variable Climate, 1999, pp. 167-195
Eds. Y.P. Abrol and Sulochana Gadgil

10. The Power of Climate Components on the Flare-up of Diseases and Pests in Rice

India is the World's second most populous nation with a population of 930 million in 1995. The country has a growth rate of 2.0 %, an infant mortality rate of 9.2% life births and a life expectancy of 59 years. Though India is still predominantly a rural country, even the urban population of over 300 million exceeds the total population of most developing nations. The mainland of India extends from 8 to 37°N latitude and from 68 to 97° E longitude. The country is bounded by the Himalayan mountain range in the North, and narrows to form the Great Indian Peninsula in the south protruding into the Indian Ocean. The Bay of Bengal is in the East of the Peninsula and the Arabian Sea is in the West. The mainland consists of four well-defined regions: i) the great mountain zone; ii) The Indo-Gangetic plain; iii) the desert region; and iv) the southern peninsula. Agriculture is the backbone of India's economy, providing direct employment to about 70 % of the rural working population. Rice is the food for 65 % of the total population in India, constituting about 43 % of the total food grain production and 47 % of total cereal production. The adoption of modern high-yielding varieties in 1965-66, increased area under cultivation, improved management practices, and widespread availability of credit to farmers, led to average annual increase of 2 % in rice production. National rice production increased fron 38.6 million tonnes in 1974-75, to 81.2 million tonnes in 1994-95. Self-sufficiency in rice was reached by 1977. The current average rice yield in India is 3t ha^{-1}. Rice yields continue to rise as a result of improved management. The high quality scented Basmati and fine-grained non-Basmati rices are also exported.

India has the largest area under rice in the World, and consequently a large diversity in rice-growing environments. Of the 42 million ha of total harvested rice area, about 45 % is irrigated, 33 % rainfed lowland, 15 % rainfed upland, and 7 % flood-prone (Prasad *et.al.*, 1995). Since the major portion (55 %) is rainfed, national production is strongly tied to the distribution of rainfall. In some of the states, erratic rainfall leads to drought during the vegetative period, but later in the season the crop may be damaged by submergence due to high rainfall. In the eastern states, damage due to flash floods can be high (Muralidharan *et. al.*, 1988). Soil fertility is often another constraint; extreme soil acidity is problem in some areas in southern and eastern India, whereas in northern India soil salinity and alkalinity can limit yield. Nitrogen, phosphorus and zinc deficiencies are also widespread. Nearly all of the rainfed area suffers from a lack of infrastructure. Input use, including nitrogen fertiliser, is very low in most rainfed rice areas. Effect of some important climatic components on the occurence of rice diseases and pests are discussed.

10.1. Prevalence of Insect Pests

In India, insect pests constitute one of the major causes of low rice yields, as the hot and humid climate required by this crop is also conducive for development and multiplication of insects. The problem has been further aggravated due to extension of areas under double or triple cropping coupled with monoculture of high-yielding, short statured, heavy tillering varieties. The modern high-yielding varieties require assured irrigation and higher doses of nitrogen fertilisers that also favour most of the insect pests. It is but natural that these changes in agroecosystem of paddy fields have resulted in significant shifts in the pattern and status of insect pests. Besides stem borer and gall midge, pests like brown planthopper, leaf folder, green leafhoppers and whorl maggot became serious occurring in more areas. These insect pests also started appearing in both *kharif* and *rabi* seasons in some of the states (Table 1). The yield losses (21 to 51 %) in rice due to insect pests were reported (Kalode, 1983) in selected states. In view of the increased pest problems and the limitations of individual methods of pest control, an integrated approach

Table 1. Incidence of insect pests in different states in India

State	SB	GM	BPH	GLH	LF	WBPH	CW	RH	STB
Andhra Pradesh	S	M	S	S	S	S	M	M	M
Assam	M		L	L	L		M	M	M
Bihar	L	L	L	L	L		L	L	L
Gujarat	L			L	L	L	L	L	L
Haryana	S				S	M	L	L	L
Himachal Pradesh	M			L	M	L		L	
Jammu & Kashmir	L			L	L				L
Karnataka	M	M	M	L	M	L	L		L
Kerala	S	M	S	L	M				L
Madhya Pradesh	M	S	M	S	M	M	L	L	L
Maharashtra	S	L	L	L	M		L	L	
Manipur		S		L	L			L	L
Meghalaya	L			L					L
Nagaland	L			L				L	L
Orissa	S	S	M	S	M	L	M	L	L
Punjab	S		M	L	S	M		M	
Rajasthan				L	M	M		M	
Tamil Nadu	S	L	M	M	S	L	L		L
Tripura	L	L		L	L			L	L
Uttar Pradesh	M	L	L	L	L	M	M	L	M
West Bengal	M	L	M	L	L			M	M

SB = Stem borer; GM = Gall midge; BPH = Brown plant hopper; GLH = Green leafhopper; LF = Leaf folder WBPH = Whitebacked planthopper; CW = Case worm; RH = Rice hispa; STB = Stink bug L = Low; M = Moderate; S = Severe based on sound ecological principles, is made for an effective pest management. Such an approach needs greater emphasis to achieve significant and sustained increase in rice production. An attempt is made here to briefly discuss the factors influencing rice pest populations. The status of pests in different states and the importance of pest monitoring that enable us to formulate suitable pest management strategies are considered.

10.1.1. Factors Influencing Pest Populations

A detailed knowledge on various factors that influence pest population would greatly help in forecasting the outbreaks of pests. But the information on rice entomological research indicates that not a single case of a completely successful forecasting technology has ever been developed, not even for a single insect on country-wide or even region-wide basis. However, research of very varied types relevant to an understanding of the insect outbreaks etc., are available from the innumerable studies on rice.

10.1.1.1. Ecological Factors

Temperature, relative humidity, rainfall and day length are the most important climatic factors affecting insect population. A particular set of these factors in a locality may have specific influence on the biology of pest resulting in an outbreak or suppression of that pest in that locality. Therefore, any generalisation made about the outbreak of pest based on a set of climatic factors always may not be correct. However, certain clues may be helpful in predicting the pest incidence.

10.1.1.1.1. Stem Borer

High temperature and low rainfall can cause severe stem borer infestation. The peak emergence of stem borer population, as indicated by light trap catches was noticed, when the maximum temperature ranged from 27.6 to 36.0° C, minimum from 19.1 to 23.3°C and relative humidity from 76.0 to 99.9. Temperatures either above 38.0° or below 15.1°C, and rainfall exceeding 158 mm/week were reported to be unfavourable for the build-up of stem borer (Kalode, 1974). Percentage incidence of stem borer dead hearts and white ears, was correlated negatively with rainfall and minimum temperature, and positively with maximum temperature (Abraham *et. al.*, 1972). Activity of stem borer was observed to be more from August to May in double cropped areas. In places having distinct generations, usually the first generation (smaller in size) appears when the plants are in the nursery or shortly after transplanting. The population increases in subsequent broods. The second or later generations of stem borer often cause serious damage. That is why stem borer is more destructive in the late planted crop or in the second crop, where double cropping is practised. During periods when there is no rice crop and the temperatures are not optimum for larval development, the full grown larvae undergo diapause. But wherever two or more rice crops are grown in a year, stem borer remains active year round. During the brief periods of non-availability of host plants, it undergoes only a temporary quiescent stage or weak diapause in the last larval instar.

10.1.1.1.2. Gall Midge

Gall midge becomes active at the onset of the monsoon, when it completes one or two generations on grasses before rice is planted and then finally transfers to the rice fields. Depending upon the climatic conditions, the peak emergence of gall midge fluctuates between July and October in *kharif*, and February and March in *rabi* season. Average maximum temperatures of 27.2 to 33.6°C, minimum of 18.0 to 24.7°C and relative humidity of 73.3 to 97 % were reported to be optimum for the peak activity of this pest (Kalode, 1974). According to Reddy (1964), the years of most serious gall midge infestations are those when early rains make the flies active, but later dry periods delay planting resulting in multiplication of the pest on grasses and migration of flies in large numbers to the late planted crop. Similarly, Rao *et. al* (1971a) reported that early rains in May and June (>30 0 mm) help the growth of grasses, ratoons and other host plants. On these food sources, gall midge completes its development and attacks the first rice crop in the season. Continuous cloudy and rainy weather prevailing from July to September coupled with high humidity was found favourable for the activity and outbreak of this pest (Hidaka and Vungsilbutr, 1971 and Rao *et. al.*, 1971a).

10.1.1.1.3. Leaf and Planthoppers

The characteristic patterns of green leafhopper build up appears to be towards the cessation of rainfall and later stages of plant growth. The light trap data from various co-ordinating centres showed that the major peaks of green leafhoppers were in *kharif* season from September to December. The maximum temperatures ranging from 26.9 to 36.0°C and minimum from 20.6 to 24.7°C and high humidity (75 to 95% RH) appear to favour this pest. Higher populations of brown planthopper were recorded from August to December in *kharif* and from April in *rabi*. The peak activity of brown planthopper was associated with a maximum temperature ranging from 28 and 36° and minimum from 19.1 to 24.7°C and high humidity. A minimum temperature below 15°C was observed to be detrimental for brown planthopper and green leafhoppers.

The abundance of leafhoppers was attributed to high temperature, low rainfall and abundant sunshine (Harukawa, 1951). According to Pathak (1968), in general, leaf and planthoppers are usually more aboundant during dry than wet season. Alam (1971) and Dyck and Hsieh (1972), however, reported high populations in the wet season.

10.1.1.1.4. Leaf Folder

In recent years, rice leaf folder has gained a major pest status in view of the severe damage caused by it in several states. Although leaf folder moths have been recorded throughout the year, they are usually more abundant during rainy season. In places with cool winters, the insect is active from May to October during which it completes four to five generations. High humidity and low temperatures appear to be important factors for its abundance (Kalode, 1984).

10.1.1.1.5. Rice Hispa

Rice hispa that was earlier supposed to be a sporadic pest in certain areas has now assumed serious proportions in many states. Outbreaks of hispa occur regularly in one or the other part of the country. Usually heavy rains in July followed by abnormally low rainfall in August and September encourages hispa to multiply rapidly. Under such abnormal monsoon conditions with fewer rainy days and more bright sunshine hours, temperatures steady along with high relative humidity. All these conducive conditions cause epidemics of rice hispa (Rao *et. al.*, 1971b; Venkatarao and Muralidharan, 1977a).

10.1.1.2. Agronomic Factors

The modern high-yielding rice varieties with luxurious growth of foliage provide a thick crop canopy beneath which the microclimate becomes quite congenial for development and multiplication of insect pests. Apparently greater severity of a particular insect pest problem is not due to any single factor. Severe insect problem may be due to a combination of many cultural or management practices. They include continuous planting of the same variety over large areas, late planting with aged seedlings, excessive use of nitrogenous fertilisers, quantum of irrigation, plant spacing and indiscriminate use of insecticides.

10.1.1.2.1 Continuous Planting

Rice is mainly grown either as a wet or dry season crop, or both. The wet season crop will have more rain and cloudiness while the dry season has less rain and more sushine. In some areas, however, two or three crops of rice are grown overlapping each other. Such type to continuous planting of the same crop, season after season favours the survival and multiplication of pests without any break and causes large scale outbreaks of insect pests. This is best illustrated by the large scale outbreaks of brown planthopper in Kerala and Andhra Pradesh where rice is grown for more than one season in a year.

Occasionally, an outbreak of hitherto unknown minor pest like Malayan black bug (*Scotinophora coarcata*) might occur (Venkatarao and Muralidharan, 1977b). Similarly, the practice of staggered planting during the season as in West Bengal also provides continuity of food for insects and enables them to multiply leading to outbreaks.

10.1.1.2.2. Excessive Use of Nitrogenous Fertilisers

It is an established fact that excess use of nitrogen frequently leads to an increase in the abundance of various pests. Israel *et. al.* (1967) observed greater gall midge incidence with increased soil application of nitrogen. Chelliah and Subramanian (1972) obtained similar results. Irrespective of the variety, when the levels of applied nitrogen increased, the intensity of gall midge infestation also increased.

Israel and Vedamoorthy (1963), Varadarajan and Nagaraja Rao (1965) and Saha and Saharia (1970) observed high stem borer damage with increased level of nitrogen. The borer damage increased with an increase in the amount of nitrogen applied. In relation to nitrogen application, there was a significant increase only in white ear damage and not in the damage from dead hearts (AICRIP, 1973-84).

Female brown planthoppers reared on plants receiving high level of nitrogen showed increased fecundity (Kalode, 1971). In plots treated with 80 lb. of N, high populations of brown planthopper were recorded (Abraham, 1957). Siginificant build-up of brown planthopper was reported by Pillai *et. al.*, (1979) on a closely planted crop at a high level of nitrogen under artificial inoculation. Dhaliwal *et. al.* (1980) observed an increase in hispa infestation in rice fields along with an increase in the applied nitrogen level (0 to 120 kg/ha).

10.1.2. Pest Monitoring and Economic Thresholds

In the ecology oriented approach to pest management, constant monitoring plays an important role. Monitoring provides information that enables to take appropriate remedial measures in time to counter the severe pest infestations and avert economic losses in grain yields. This is mainly concerned with know-how on deriving the time when the population will reach the economic threshold level. Due consideration is also given to all the pest suppressing factors such as plant resistance, natural enemies and adverse physical factors while arriving at threshold levels. The usefulness of pest monitoring can be best illustrated by the example of brown planthopper epidemics, that occurred in Kerala during 1974. Initially, farmers could not identify the pest because brown planthopper feeds at the basal portion of plants. But in later years, due to the efforts put in by the state government in monitoring, the pest was brought under control and huge losses were averted.

The level of pest population is a very important consideration for taking up control measures. Populations must be maintained at levels below those causing economic injury. The economic threshold indicates the pest density at which control measures should be taken in order to prevent the increasing pest population from reaching the economic injury level. The economic threshold rather includes cost of control, crop value, changing susceptibility of crop to insect damage, changing density and composition of pest populations value. The economic threshold values were tentatively proposed for major pests of rice (Kalode, 1984). These threshold values (Table. 2) could be modified as per the local situation.

Table 2. Economic threshold levels as decision aids for intervention in management

Insect pest	Threshold level
Brown planthopper	5-10 insects/hill
Green leafhoppers	2 insects/hill in tungro endemic areas or 20 insects/hill in other areas.
Stem borer	5 % dead hearts in vegetative stage or 1 moth/m^2 in flowering stage.
Gall midge	1 gall/m^2 in endemic area or 5 % galls during tillering phase
Rice hispa	1 adult or 1-2 freshly damaged leaves/hill
Leaf folder	1 to 2 freshly damaged leaves/hill
Thrips	5 % plants with rolled leaves.
Whorl maggot	20 % damaged hills.
Case worm	1-2 cases/hill
Cut worm	1 larva/hill
Stink bug	5 or more bugs/m^2

Careful monitoring of pest abundance and the threshold values would greatly help to reduce the pesticide application which in turn would enable to maintain the ecosystem in proper balance.

Biological control is the most important component of the integrated pest management strategy. Many natural enemies, predators, and parasites on insect pests and disease causing organisms are known to occur widely in rice ecosystems. The success of using any biological control agent would largely depend on understanding the interaction of the seasonal climatology of the place of introduction.

10.2. Rice Diseases in Field

Diseases are often hurdles in increasing per unit rice yields. Under different rice ecosystems, the important of diseases in rice production has been assessed (Table. 3). The rice blast, a disease known since ages, is induced by *Pyicularia grisea*. Blast continues to be of highest significance across rice ecosystems. In the past, blast, brown spot (*Helminthosporium oryzae*) and stem rot (*Sclerotium oryzae*) were the serious diseases. Consequent to the spread of high-yielding varieties and new production practices during 1970s, diseases like bacterial leaf blight (*Xanthomonas oryzae pv.oryzae*), sheath blight (*Rhizoctonia solani*), sheath rot (*Sarocladium oryzae*), tungro virus (transmitted by *Nephotettix* spp.) and bacterial leaf streak (*X. transluscens pv oryzae*) have gained importance over the traditionally known diseases, especially stem rot and brown spot. False smut (*Ustilaginoidea virens*) and discoloration of rice grain caused by several fungi have been of minor significance with occasional concern in certain regions. Rice diseases induce losses to the tune of 15 to 20 % annually.

Table 3. The level of importance of diseases in different rice ecosystems in India.

Disease	Rainfed uplands	Irrigated areas	Rainfed shallow lowlands (0-30 cm)	Rainfed intermediate lowlands (30-50 cm)	Semideep water areas (50-100 cm)	Deep water areas (> 100 cm)
Blast	35	25	25	20	15	15
Sheath blight	15	15	10	15	20	25
Brown spot	20	*	10	*	*	*
Sheath rot	10	5	10	*	*	*
Stem rot	*	5	10	10	15	40
False Smut	*	5	5	5	*	*
Grain Discoloration	10	*	10	10	10	*
Bacterial leaf blight	*	20	20	20	20	20
Bacterial leaf streak	10	5	10	*	*	*
Tungro	*	20	10	20	20	10

*These diseases are not of serious concern in respective ecosystems.

10.2.1. Rice Diseases and Climate

Climate and weather not only influence the growth of rice crop, but also selectively induce a specific disease to develop. On the basis of climate influence on the growth, diseases of rice can be grouped into two categories, monocyclic and polycylic. The monocyclic diseases such as, stem rot, sheath rot, and false smut are less influenced by the ambient weather conditions. Epidemics of monocyclic diseases are relatively rare events in the sense of an explosive increase of their population. In contrast, the polycyclic diseases such as, blast, brown spot, bacterial leaf blight and rice tungro virus that invade the aerial parts of the plants are subjected to constant interaction with weather. Therefore, they easily attain epidemic proportions to cause heavy losses. In these diseases, the epidemic growth is highly influenced by the weather in the following ways:

. Direct effect on pathogen,

. Direct effect on host,

. Direct effect on the interaction between host and pathogen,

. Direct effect on the interaction between host and pathogen at given time and space.

Valuable information is available on the influence of weather and climate on diseases of rice (Padmanabhan, 1974; Ou, 1985; Narain, 1988; Sharma and Mehrotra, 1988; Nagarajan and Muralidharan 1995). However, the data gathered on diseases other than blast remain un-utilized for developing growth models of diseases or prediction or forewarning systems. The major environmental factors that influence the onset and outbreak of diseases have been critically discussed earlier (Mayee, 1995).

10.2.2. Role of Weather in Epidemic Development

Brown spot disease of rice has been responsible for the Great Bengal Famine (Padmanabhan, 1973). This disease is favoured by a temperature range of 23 to 25°C and relative humidity about 80%. The conidial dissemination is affected by wind current. Rainy days followed by days with bright sunshine contribute to the spread of infection. The number of sunshine hours influences the size of lesions in susceptible varieties. In the air, spore counts reach high levels during November and maximum spores are available between 0300-0900 hour, in the early morning. The lesion size increases when the leaf wetness period is prolonged at temperatures ranging from 20-25°C. Associated with these weather factors, brown spot epidemic is nearly ensured, if the soils are nutritionally poor.

In contrast, blast epidemics occur in host population nourished on highly nutritive soils, particularly in soils rich in nitrogen. The prime factors associated with blast epidemics are high levels of nitrogen application, susceptibility of cultivar and favourable weather. For example, leaf wetness of 10 hour or more per day in conjunction with high rainfall or dew deposition, and low night temperature around 20 to 26°C and relative humidity of >90 %, are extremely critical for blast outbreak in fields.

The sclerotia of the sheath blight fungus survive in soil and the pathogen produces basidiospores that cause leaf and panicle infection. They are readily disseminated in the wind. High humidity favours vertical and horizontal development of the disease.

Factors responsible for rapid development of bacterial leaf blight include high varietal susceptibility, high N application, continuous drizzle and high humidity. The amount of rainfall (> 300 mm) and distribution patterns (12-15 days per month) in July- September primarily determine the rapidity of the growth of bacterial leaf blight pathogen. The wet season temperatures in the range of 23- 31°C, are conducive to disease outbreak. For wilt phase in seedling, called *kresek*, the temperatures in the range of 28-35°C, are favourable when early inoculum is available.

Rice tungro virus disease is very destructive in several rice growing areas of the country. Epidemics of tungro were frequently encountered in many rice growing regions (Anjaneyulu *et. al.*, 1982). The virus is transmitted in semi-persistent manner by several species of the vector, green leaf hopper. *Nephotettix* spp. Temperature, humidity, wind velocity and sunshine hours influence both the vector and virus. Tungro virus disease occur in epidemic form in a cyclic pattern. Coincidence of vector population, virus inoculum, susceptible young foliage along with prevalence of high humidity and temperatures in the range of 25° - 30°C, are necessary for outbreaks to occur.

10.2.3 Weather Based Prediction of Blast in Upland Paddy Region

Marathwada region of Maharashtra is famous for growing a typical rice culture referred to as upland paddy under rainfed as well as irrigated situations. The tall and lodging- susceptible cultivars, Ambemohar and Tujiapur, have now been replaced with semi-dwarf non-lodging cultivars, Prabhavati, Sugandha and Parag that are also blast-susceptible. Blast is the major disease problem in the area and management is based on chemical control (Jadhav *et. al.*, 1986).

During *kharif*, farmers grow only susceptible varieties and the humidity in the crop growth period remains < 80 %. All the cultural practices followed are nearly similar including the use of 60 kg N/ha fertiliser in three split doses. Blast severity prediction was made on the basis of nycto-temperature (prevalence of 20 - 26°C) data during July - September period. Initially blast severity was computed as the number of spots on all the leaves of 10 plants selected at random in the fields and correlated with the nycto-temperature regimes to enable predictions (Table 4). Using this method, predictions were made three times. Simultaneously, actual data on disease incidence were gathered for comparison. All the predictions were rated into three categories using 0-9 scale; 0 as totally wrong and 9 as totally correct. In addition, using the following equation, the rates of spread of infection (Vanderplank, 1963) on the plants were assessed through periodic observations:

$$r = 2.3/(t_2 - t_1) \log \{x_2/(1 - x_2) - x_1/(1 - x_1)\}$$

Table 4. Prediction of blast severity based on nycto-temperature regimes (20° to 26°C).

Favourable period duration	Crop stage	Blast severity (number of leaf spots or neck blast-NB%/plant)
1. Continuous week	Tillering Panicle intitiation After flowering	> 10 25-35
2. 4 days in a week	Tillering Panicle initiation After flowering	> 10 % 5-10 20-25
3. 1 day in a week	Tillering Panicle intiation After flowering	5-10 % < 1 5-10 < 1 %

where r = infection rate in units per day, x_1 and x_2 are disease severity on 0-9 scale at time t_1 and t_2, respectively.

Of the predictions (Table 5) made during tillering phase, only in five years, the predictions on blast matched the actual disease occurrence; in two years, the predictions were bad; and in one year,

Table 5. Prediction of leaf blast at tillering stage in upland rice.

Year	Predicted severity (spots/plant)	Actual severity (spots/plant)	Correctness of prediction (scale*)
1982	10-12	8-10	8.2 G
1983	5-10	1-2	2.3 B
1984	5-10	5-7	8.0 G
1985	15-18	12-15	8.0 G
1986	7-10	11-13	8.9 G
1987	1-2	12-15	1.3 B
1988	5-8	15-18	4.5 M
1989	10-14	12-15	8.7 G

0 to 9 scale : 0 is totally wrong and 9 is totally correct prediction, qualified as BAD (B): 1-3, MODEST (M): 4-6 and GOOD (G): 7-9.

the prediction was only partially correct. At panicle initiation stage, only three predictions on the leaf blast matched with the actual disease severity in the fields (Table 6). Predictions made on neck blast severity in upland rice matched with the actual data gathered in the fields (Table 10.7) in three years; predictions were totally bad in five years. During early crop growth stage, 63 % of predictions made were

Table 6. Predication of leaf blast at panicle initiation stage in upland rice.

Year	Predicted severity (spots/plant)	Actual severity (spots/plant)	Correctness of prediction (scale*)
1982	22-25	12-15	4.5 M
1983	12-15	2-5	2.0 B
1984	12-14	14-16	8.5 G
1985	25-30	24-26	8.5 G
1986	20-22	18-20	8.6 G
1987	15-18	28-32	1.9 B
1988	10-15	20-22	4.0 M
1989	25-28	8-10	2.9 B

Table 7. Prediction of neck blast at panicle development stage in upland rice.

Year	Predicted severity (%)	Actual severity (%)	Correctness of prediction (scale*)
1982	8-10	2-3	2.5 B
1983	2-4	7-8	1.8 B
1984	2-3	3-4	8.7 G
1985	5-7	8-9	8.5 G
1986	8-9	7-8	8.6 G
1987	6-8	3-4	2.8 B
1988	7-8	3-4	1.3 B
1989	1-2	6-7	3.0 B

found to be correct. This level of correctness declined to 38 % in predictions made at later crop growth stages. Neither the onset of monsoon nor the date of sowing drilled paddy showed any influence on the disease occurrence. Only when there was a uniform progress in the infection and spread of the disease, the predictions were found to be correct. In the blast disease progress, in the years when predictions proved bad, variable rates of infection, were observed (Table 8). Therefore, a single factor like the nyctotemperature is not enough to derive accurate prediction on the disease in incidence.

Table 8. The infection rate and blast disease progress in farmers field during eight years and predictability

Description	Periodical infection				
	rate (r) at tillering	r/day at panicle initiation	r/day at flowering	Observed in year	Analysis of prediction
Uniform high progress rate	0.22	0.24	0.20	1985, 1986	Good
Initially high subsequent decline	0.24	0.10	0.09	1982, 1987, 1988	Bad
Initially low subsequent high	0.05	0.18	0.25	1983, 1989	Bad
Uniform low progress rate	0.05	0.08	0.05	1984	Good

10.2.4. Blast Prediction-simulation Model

Among the major fungal diseases of rice, blast caused by *Pyricularia oryzae* syn. P. grisea (= *Magnaporthe grisea*) is a major constraint in rice production as it causes severe yield losses from 40-90 %. Several factors are responsible for the outbreak of blast. The pathogen attacks rice crop at all growth stages. The disease incidence also varies with the region, season, cultivar and virulence of the pathogen. Hashioka (1965), Suzkui (1975) and others have comprehensively discussed the role of meteorological factors in the epidemiology of rice blast disease. Several simulation techniques have been used to forecast epidemics of blast (Kim and Kim, 1991 and Teng *et, al.*, 1991).Manibhushanrao and Krishnan (1991) designed a simulator, EPIBLA for forecasting leaf blast progress. This model is based on BLIGHTAS, which is a prediction model developed in Japan for rice sheath blight disease. EPIBLA has limited application and predicts pathogen spore concentration in air and with its aid predicts the progress of disease over 7-d period. Manibhushanrao (1994) has reviewed the blast literature on modelling and forecasting system.

The severe outbreak of blast disease generally occurs under the favourable weather conditions. The disease severity was also influenced by the intensive cultivation of semi-dwarf, high-yielding and high N responsive rice varieties. Also, soil nutrition (N-fertilisation) and cultural practices contribute to spread and intensity of the disease.

Attention has been given to developing simulation models using systematic approaches including statistical analysis. The forecasting system was developed by continuous monitoring of weather parameters and exact quantitative information on the blast pathogen's air-borne conidia.

The most common climate conditions that favour the rice blast epidemics are high humidity, lack of sunshine and long periods of drizzle rain (Suzuki, 1975). However, Yamanaka (1965) reported that temperature was the most important single determinant of blast disease and rain could be a limiting factor. In our investigations, dew and high relative humidity (RH) contributed more for spore release, while low RH and maximum temperature reduced the amount of spore release. Further, the spore releasing capacity differed between various types of rice blast lesions investigated. Spore liberation was maximum in purple grey (pg) coloured lesions than yellow purple grey (ypg), yellow brown grey (ybg), brown grey (bg) and brown (b) lesions. These results confirm the earlier observations of Dhua (1989). The area of the lesion (10 mm^2 was uniformly used. The pg lesion alone discharged respectively, 11-and 200-folds more number of spores than bg and b lesions. The pg lesions were reported to be typical of most susceptible reaction type (Abumiya, 1955). The susceptibility decreased when the margin colour of the lesion changed, from dark purple to brown which is reported to be the indication of the resistant reaction (Suzuki, 1975).

Under temperate conditions, Suzuki (1969) reported the diurnal periodicity of spore release at 1800 to 1900 h followed by 0400 to 0500 h. In the study, maximum spores were observed at 0400 and 0800 h, and this might be due to the prevalence of favourable microclimate (more than 90 % RH) from 2000 h.

The blast pathogen's spore number was estimated by multiple regression analysis, which fitted more closely to observed values. Under temperate conditions, Kiyosawa (1972) reported that exponential model fitted well with the observed data. Further, multiple regression analysis has been used especially between blast disease and other independent or concomitant variables (Ono, 1965). However, in our study, the disease incidence was extremely low and consequently the differences in yield amongst cultivars were negligible. Further studies on the systematic recording of the climatic factors, monitoring the spore release and understanding the interactions of various control measures are warranted to develop on the effective of forecasting of rice blast disease.

10.3. Disease Prevalence

In most rice growing areas, every year many diseases occur at varying intensities and cause economic losses. The Directorate of Rice Research has been organising surveillance routinely from 1965, with the help of co-operating scientists at the different Agricultural

Universities and Extension personnel at the Departments of Agriculture in the various states. The production oriented surveys (DRR, 1975-97) and basic research on the epidemiology of rice diseases have renabled us to understand the outbreaks of diseases in rice. The current status of rice diseases in different states is shown in Table 9.

Table 9. Incidence of diseases in different states in India.

States	Fungus										Bacterium			Virus
	Bl	Nbl	Shbl	FS	ShR	BS	FR	LS	SR	Ud	Kr	BLB	BLS	RTV
Andhra	S	M	M	S	L			L				S		L
Assam	S	L	M	L			L	L				S		
Bihar	M	L		S	L	L						L	L	
Gujarat	M	L		M	L							M		
Haryana	M	S		M	L	L			L		L	S		
H.P.	S	S			M	M						L		
J & K	S	S	L		M	M						L		
Karnataka	S	M		L	L	L						L		
Kerala	L		L	L	L	L				M		M		L
M.P.	M			L	L	L			L			M	L	
Maharashtra	M		L	M								S		
Manipur	S	S												
Meghalaya	M	M										M		
Nagaland	L	L										L		
Orissa	M					M							S	L
Pondicherry	M	M	L		L									
Punjab	M	M		L		S					S	S		L
Rajasthan	L					L						L		
Tamil Nadu	S	M	L	L	L	L								
Uttar Pradesh	M	M	L	L		L					L	M		M
West Bengal	M	L	L	L	L	M						L		M

Bl = Blast; Nbl = Neck blast; ShBl = Sheath blight; FS = False smut; ShR = Sheath rot; LS = Leaf scald; SR = Stem rot; Ud = Udbatta; Kr = Kresek; BLB = Bacterial leaf blight; BLS = Bacterial leaf streak; RTV = Rice tungro virus.

10.3.1. Factors Influencing the Outbreak of Diseases

An epidemic outbreak of blast disease damaged severely the entire rice crop in Thanjavur delta of Tamil Nadu in 1919 (Padmanabhan, 1965). During 1950s, blast became less severe (Padwick, 1950).

Although *Pyricularia grisea* has been known for a long time, with the introduction of high-yielding varieties in 1967-68, blast became one of the major limiting factors in rice production. Severe blast damage became a common occurrence, particularly in upland and hill ecosystems. In the last decade, blast disease is appearing more regularly in certain districts in Tamil Nadu and Andhra Pradesh in the irrigated and rainfed shallow lowland rice ecosystems. The yield loss from blast disease was estimated at 1 % of the total production in Andhra Pradesh. In endemic areas, depending on the severity of the epidemic, losses in yields ranging from 20-80 % have been reported (Muralidharan and Venkatarao, 1979a, 1987). Severe damages by leaf and neck blast on Basmati rices have been estimated in Punjab and Haryana (Ahuja, 1988). At high elevations in Jammu and Kashmir, western Uttar Pradesh and Himachal Pradesh, and in Assam, blast has been a major constraint to increasing rice production (DRR, 1975-97). In the hill districts of Coorg, Chigmagalur and Shimoga in Karnataka, Intan variety succumbed to severe neck blast. In Tamil Nadu, during 1983-84, as a result of wide spread blast epidemics or IR 50 in North Arcot, South Arcot, Chingleput, Pudukottai, Madurai, Periyar and Coimbatore districts, crop was lost by the farming community. Similar outbreaks of blast in Tamil Nadu and Andhra Pradesh have led the farmers to abandon large scale cultivation of IR 50. In Kerala, the entire dry sown crop, which is roughly equivalent to 70 % of Virippu season rice, was affected by leaf blast resulting in considerable loss. The intensity of occurrence of rice blast in different rice growing states in India has been estimated (Table 9).

Many factors are held responsible for the outbreak of blast disease epidemics. The weather conditions, particularly low night temperature regimes and moisture availability for the pathogen to grow and multiply, play a decisive role in the development of blast. The number of days with minimum temperatures of 20°C was suggested to predispose plants to blast disease. Sadasivan *et. al.* (1965) found that rice plants on exposure to low nycto-temperature for several days became susceptible to infection by *P. grisea*. Humidity or rainfall without any such predisposition did not increase the susceptibility. Low nycto-temperatures lead to exaggerated nitrogen metabolism that was considered to increase susceptibility. The occurrence of minimum temperature ranging from 20-25°C along with relative humidity of > 95% for a week or more during any susceptible crop growth (seedling, tillering and panicle initiation) stages, was found to be associated with blast epidemic (Padmanabhan *et. al.*, 1971). Age of the plants, time of planting (Venkatarao and Muralidharan 1982a), density of plant population in the fields (Venkatarao and Muralidharan, 1982b), and nitrogen application (Amin and Venkatarao, 1979) are well known to interact with climate components and alter the intensity of blast incidence. Muralidharan and

Venkatarao (1980a) developed a simple method to predict blast outbreak using all this information. They used data on disease incidence in trap plot nursery, in conjunction with favourable meteorological conditions. Later an explanatory empirical model was developed (Muralidharan and Venkatarao, 1987; Nagaragan and Muralidharan, 1995) with simple rules to predict outbreaks of blast:

For leaf blast outbreak:

(1) Moderate high incidence of leaf blast in trap nurseries with any locally susceptible cultivar sown at weekly interval, and 2) continued prevalence of low night temperatures (< 20°C) coupled with high relative humidity (> 90%) for 7 to 10 days, or cloudiness, dew fall or light drizzle.

For neck blast outbreak:

(1) A low to moderate leaf blast incidence on crop during tillering stage, or at least in trap nurseries, and 2) low night temperatures (< 20°C) during panicle initiation to flowering stage, or cloudiness, dew fall or light drizzle during the reproductive phase.

10.3.1.1. Sheath Blight

Sheath blight disease caused by *Rhizoctonia solani* was first reported from Gurdaspur district (Pracer and Chahal, 1963) Punjab. Later this disease was found to damage rice in Uttar Pradesh (Kohli, 1966), Kerala, Andhra Pradesh, Tamil Nadu, and Jammu and Kashmir (Rao and Kannaiyan, 1973, Amin, 1975). In the last decade, the disease has become widespread.

In Andhra Pradesh, sheath blight is a constraint in coastal districts (Subbaiah, 1986, Venkatarao *et. al.*, 1988). Although the disease has been recorded in many districts in Tamil Nadu, it is persistent in the soils of certain areas, particularly in Thanjavur and Tirunelveli districts. In Virippu crop with high plant population density per unit area, leaf infection by *R. Solani* is a common feature in Kerala. Sheath blight infection was severe in clay soils, but was less in red, alluvial, laterite and sandy soils. Infection may occur in the temperatures range of 23-35°C accompanied by high relative humidity (> 90%). When temperatures range from 28 to 30°C, the fungus produces abundant sclerotia on plants in the field. Plants that are susceptible to brown plant hoppers were more prone to sheath blight pathogen attack (Dath, 1990).

In Punjab, indiscriminate use of nitrogenous fertiliser, continuous cultivation of susceptible cultivars over a wide area and excessive irrigation have probably facilitated the establishment of sheath blight disease firmly. In north India, the disease is known in rainfed uplands as banded blight. In south India, the disease is usually found at late crop growth stages. During this period, the generally prevalent high humidity and temperatures favour horizontal and vertical spread of the disease. The vertical spread drastically reduces the yields. Saxena and Chaubey (1972) reported the abundant occurrence of *Thantophorus cucumeris*, the perfect stage of the pathogen. The basidiospores incite wide spread leaf and panicle infection leading to rapid spread of the disease. The pathogen is polyphagus. Innumerable sclerotia produced by the pathogen help in over wintering and perpetuation of the disease. The inoculum is readily available in irrigated and rainfed lowland ecosystems. Host resistance to *R. solani* is poor in rice varieties. Therefore, under high nitrogen fertiliser application and plant density, presence of adequate moisture is enough to trigger the spread of the disease. Fortunately, sheath blight incidence is usually not homogenous probably due to the predominant saprophytic ability of the fungus. Total collapse of plants after flowering as a result of severe infection is always restricted to patches, or at best a few small fields.

10.3.1.2. Sheath Rot

Sheath rot disease caused by *Acrocylindrium oryzae* markedly reduced rice productivity (Venkataraman and Muralidharan, 1988). The intensity of the disease was severe whenever panicle emergence was affected by climatic factors. The fluctuations in monsoon and the late rains, particularly from panicle initiation to booting stages in *kharif*, enable sheath rot disease to appear and cause damage to production in many states. The intensity of sheath rot is enhanced in crops already infected with other biotic stresses such as rice tungro virus, bacterial blight or stem borer (Reddy and Ghosh, 1985; Venkataraman *et. al.*, 1987). Besides reducing grain yields, the disease has been found to affect grain quality (Ahmed *et. al.*, 1978). Sheath rot, considered to be a minor disease, has assumed greater importance in the recent years, particularly in Uttar Pradesh, Punjab, Karnataka, Bihar and West Bengal.

A. oryzae is known to infect many host species. Some information is available on the favourable factors for the development of this disease. The primary factor responsible for sheath rot outbreak is the availability of moisture at the site of infection. Occurrence of high relative humidity as a result of rain, drizzle or dew fall at flowering stage, helps sheath rot pathogen to establish and spread (Muralidharan and Venkatarao, 1980b).

10.3.1.3. False Smut

False smut disease of rice was considered as a minor disease for more than a century. In 1980s, however, false smut gained importance, partly due to adoption of intensive production technology and changed cultural practices. Rainfall during panicle initiation to flowering period favours disease development. As the number of rainy days increased, there is a direct increase in the disease severity (Muralidharan, 1988). Severe outbreaks occurred during 1983 and 1985 in Andhra Pradesh, Bihar and Uttar Pradesh (Muralidharan *et. al.*, 1990). The favourable conditions for rapid false smut disease development on rice are known.

10.3.1.4. Foot Rot Disease

Foot rot disease affects seedlings in Assam and a few other rice growing areas in the north India. The infected seedlings rot and die in the nursery. The pathogen, *Fusarium moniliforme* survives in plant debris and soil as macroconidia and sclerotia. The causal organism is known to produce growth promoting substances as a result of which infected plants emerge well above other healthy plants. The favourable factors for disease development have been studied.

10.1.3.5. Brown Spot

In India, brown spot occurs every year in relatively mild form in both upland and lowland rice. Occasionally, in patches, the disease incidence might be severe. Brown spot is widely prevalent in Eastern India and along coastal rice growing areas. The source of inoculum comes from the infected plant debris, straw and seeds. Air-borne conidia on deposition over leaf surface, cause brown spot infection and secondary spread of the disease. As infection occurs late in the season, the damage is negligible (Venkataraman *et. al.*, 1990).

10.3.1.6. Bacterial Blight

The first record of this disease in India was from Maharashtra (Srinivasan *et. al.*, 1959) where bacterial leaf blight incidence was destructive since 1951. The disease was considered to be localised to Maharashtra, until an outbreak occurred in Shahbad district of Bihar in 1963. This epidemic was identified and proved to be bacterial leaf blight caused by *Xanthomonas oryzae* (Srivastava and Rao, 1963a) In 1965, a high-yielding cultivar Taichung Native 1 was tried in farmers' fields at several locations in India. Despite subnormal rainfall, bacterial leaf blight appeared in epidemic form on TN1 in most of the fields. Further expansion of area under TN1 in India led to devastating epidemics of bacterial leaf blight in several states. Thus with the introduction of a susceptible variety, bacterial leaf blight disease that was considered to be a minor disease, assumed national importance following wide spread occurrence of epidemics in just two years.

The disease is usually more prevalent during both south-west and north-east monsoon periods. The fluctuations in the intensity of occurrence of bacterial leaf blight in India are known (Table 10.9). Moderate to severe incidence of this disease, occurs during winter period, particularly in double cropped areas of Andhra Pradesh, Maharashtra and Tamil Nadu.

Bacterial leaf blight has been shown to be a seed-borne disease (Srivastava and Rao, 1963b). As the disease spreads through irrigation water, large scale cultivation of a susceptible variety poses a constraint to production. In the absence of chemical control measures, the disease continues to threaten productivity in many states. The pathogen causes typical vascular wilt. Therefore, bacterial blight disease is not amenable to management at field level. In Punjab, Haryana and western Uttar Pradesh, *kresek* or seedling blight causes more damage in nursery. *Kresek* is followed by leaf blight at later stages of crop growth resulting in considerable yield losses in most of the popular rice varieties. During August-September, prevalence of continuous drizzle and high humid condition enable rapid development of leaf blight disease in rice. The amount of rainfall (> 300 mm) and its distribution pattern (12-15 days per month) in July-September primarily determines the degree of disease development in the country. For *kresek* development, a range of minimum-maximum temperature of 28-35°C is required (Reddy, 1983). The minimum-maximum temperature range of 23-31°C normally prevalent in the wet season is conducive for blight pathogen to grow and spread. Cyclone winds and intermittent rainfall can induce an outbreak of leaf blight in most coastal rice areas (Muralidharan and Venkatarao, 1979b).

10.3.1.7. Rice Tungro Virus

With the introduction of high-yielding varieties in India, yellowing disease on rice was observed at a few locations. Viral infection in most cases was suspected. However, only a systematic study in parts of Delhi and Eastern India (Orissa, West Bengal and Bihar) led to the identification and confirmation of the existence of rice tungro virus in our country. During 1969, an epidemic outbreak of rice tungro virus disease appeared in parts of east Uttar Pradesh, Bihar and West Bengal. Later epidemics re-occurred in 1973 and 1981 in the Eastern India. However, such severe outbreak were discontinuous through time and space. In the last decade, tungro epidemics have become widely prevalent in peninsular India. Tungro epiphytotic occurred in 1979 at Thanjavur and in 1982 at Periyar and North Arcot districts. A sporadic outbreak occurred in Chengalpet district in 1983; but in 1984, tungro epidemic caused considerable loss. Similar outbreaks during 1983-84 occurred in Ranga Reddy, Nellore and Chittor districts in Andhra Pradesh (Ghosh, 1985; Shanmugam and Jayaraj 1988). When the crop seasons overlap and excessive uses of fertilisers are made in cultivation of high-yielding susceptible varieties, the vectors, namely, *Nephotettix virescens* and *N. nigropictus* to rapidly multiply and transmit the disease (Siddiq *et. al.*, 1994). Over the last two decades, there has been a gradual shift from the original locus of Eastern India, to new areas in peninsular India. Although tungro incidence was correlated with field populations of the vector *N. virescens*, it has not been possible to trace the course of the flight of the leaf hoppers.

Evidence gathered does not indicate any direct correlation between tungro infection and meteorological conditions. However, in tungro epidemic years, large green leaf hopper populations were recorded. Cool and moist climate favours the build-up of the vector population. The most favourable period for the vector is September-October. Often early monsoon rains help vector to build-up population levels. Availability of virus inoculum, high population or vector, and susceptible crop lead to severe outbreak of tungro disease in rice fields (Anjaneyulu *et. al.*, 1986). The untimely occurrence of typhoon in May 1990 (peak summer) enabled survival of green leaf hoppers along with the virus. The availability of rice plants in irrigated ecosystem more or less throughout the year makes the vector and the virus to multiply (A. Ghosh, personal communication). This ultimately led to a severe epidemic of tungro, in 1,00,000 ha during 1990 wet season.

Transmission of tungro virus from plant to plant depends on the type of virus particle present in the diseased plants that serve as a source of inoculum. While rice tungro spherical particles can be transmitted independently, transmission of bacillus particles by the vector depends on the availability of spherical particles. Vectors, therefore, may or may not transmit rice tungro bacillus particles that cause rapid disease development in infected plants (Siddiq *et. al.*, 1994). Surveillance through light traps, mobile nurseries to identify vector population, and inoculum estimation with ELISA test to index viruliferous green leaf hoppers, was suggested to enable a forecast on tungro outbreak (Anjaneyulu, 1994).

10.3.2. Prediction Using Survey and Threshold Levels

In disease management, setting priority through identification of the key pathogens in the given target area is the first step. The quantitative estimates on the contribution of each constraint to the total loss provide a rational base for arriving at a decision for intervention. Threshold levels for various diseases (Table 10) as decision aids for an intervention in management, have been delineated. Characterisation of explanatory variables that cause host-parasite interaction, such as variety planted, yield, etc., help in detecting the patterns of relationships among variables. Further clues are obtained from a conceptual qualitative model of host-parasite system constructed using information on the driving variables (Table 10) and on the components of monocyclic infection (Table 11). A model can be developed with these parameters for use as a predictive tool in individual environments for their potential to favour a particular disease. The various components of monocyclic infection are affected by a set of climatic, edaphic and cultural factors. All disease epidemics can be explained by three basic phenomena-latency, infectious period and a multiplication factor. Any additional details would merely be expansions of the three variables describing these phenomena (Venderplank, 1963). The epidemics increase in space and time causing yield losses. Within field movement and dispersal gradients and large area movement or migration provide more information on the nature of the epidemic outbreak and help predictions. Such forecasts enable crop planning, deployment of resistant varieties, rescheduling planting date, harvest date and sequence crops to effectively alter the recurrence of epidemics.

Table 10. Complexity in the driving variables in disease development

Source of initial inoculum
Amount of initial inoculum
Single point or multipoint primary infection
Reproductive capacity of the pathogen
Dispersal characteristics
Infection site wetness
Temperature regimes
Rainfall
Dew deposition and duration
Relative humidity
Solar radiation
Wind speed and direction
Fertility status of soil and plants
Host phenology
Host morphology
Host condition-vulnerability
Host genotype

Table 11. Components of monocyclic infection

Germination	
Appressorium formation	
Penetration	Entry
Latency	
Sporulation	Multiplication
Lesion growth	Tissue or organ invasion
Spore dispersal	
Spore Survival	

10.4. Future Prospects

Problems caused by diseases and insect pests in rice production system are complex. Pathogens or pests attack, the damage caused and the ways in which farmers, extension specialists, pesticide dealers, researchers, administrators and politicians deal with them are influenced by a range of factors. The most important factors that govern action from farmer include his socio-economic condition and availability of resource to meet an exigency arising from pest attack (Muralidharan *et. al.*, 1991). Weather anomaly can often drastically affect epidemic outbreak. Use of external resources in rice production, especially nitrogenous fertilisers and pesticides, is unprecedented and therefore historical experience is valueless. In order to conserve natural resources and economise pesticide use, prediction of an outbreak of disease or pest attack is extremely useful. Today we have basic information on the biology of causal organisms, weather influence on host and pathogens or insect pests, and on the modern crop production practices. To develop decision support, all these data may be combined to predict the outcome of available management options. Comprehensive models are usually inappropriate for prediction due to their large input requirement. Summary models that capture the essential components of system dynamics hold the key to success in forecasting a disease or insect pest's outbreak in India.

10.5. References

ABRAHAM. C.C. THOMAS. B., KARUNAKARAN. K. and GOPALKRISHNAN, R. 1972. Effect of planting seasons and the associated weather conditions on the incidence of the rice stem borer, *Tryporyza incertulas* (Walker). *Agric. Res. J.* Kerala. **19**(2): 141-151.

ABRAHAM, E.V. 1957. A note on the influence of manuring on the incidence of Fulgorid *Nilaparvata lugens* Stal on paddy. *Madras Agric. J.* **44**: 529-532.

ABUMIYAH, H. 1955. On the infection types of leaf blast disease of rice plant. Jubilee Pub. Commem. 60th Birthday of Profs. T. Tochinai and T. Fukushi. Spporo, Japan, pp. 197-201.

AHMED, N.J., KANNAIYAN, S. and VENKATRAO, A. 1978. First record of *Acrocylindrium* sheath rot disease of rice in India. *Sci. & Cult.* **75**: 544-545.

AHUJA, S.C. 1988. Current status of blast and sheath blight of paddy in Punjab and Haryana. In: Proc. Rice Pathology Group Discussions. Directorate of Rice Research, Hyderabad. (Mimeographed)

AICRIP. 1973-84. Annual Reports. All India Co-ordinated Rice Improvement Project, Directorate of Rice Research, Hyderabad-500030, India.

ALAM, S. 1971. Population dynamics of the common leafhopper and planthopper pests of rice. Cornell University, Ph.D. thesis, p. 141.

AMIN, K.S. 1975. An improved method for evaluating rice sheath blight. *Phytopathology*. **65**: 214-215.

AMIN, K.S. and VENKATARAO, G. 1979. Rice blast control by nitrogen management. *Phytopath. Z.* **96**: 140-145.

ANJANEYULU, A. 1986. Virus diseases of rice India. *Trop. Agr. Res. Ser.* 19. Tropical Agricultural Research Center, Tsukuba, Japan. pp. 14-19.

ANJANEYULU, A., SATHPATHY, M.K. and SHUKLA, V.D. 1994. Rice Tungro Virus. Oxford & IBH Publishers, New Delhi. p. 228.

ANJANEYULU, A., SINGH, S.K. and SHENOY, M.M. 1982. Evaluation of rice varieties for tungro resistance by field screening technique. *Trop. Pest. Manage.* **28**: 147-155.

CHELLIAH, S. and SUBRAMANIAN, A. 1972. Influence of nitrogen fertilisation on the infestation by the gall midge *Pachydiplosis oryzae* Wood-Mason, Mani in certain rice varieties. *Indian J. Entomol.* **34**: 255-266.

DATH, A.P. 1990. Sheath Blight Disease of Rice and its Management Associated Publishing Co., New Delhi. p. 129.

DHALIWAL, G.S., SHAHI, H.N., GILL, P.S. and MASKINA, M.S. 1980. Influence of nitrogen levels on rice hispa incidence. *Inter. Rice Res. Newsl.* **5**(1): 14-15.

DHUA, U. 1989. Relation between size, colour and spore liberation from rice blast lesion. *Phytopath. Z.* **124**: 27-30.

DRR. 1975-97. Production Oriented Survey. Directorate of Rice Research (AICRIP), Hyderabad, India.

DYCK, V.A. and HSIEH, C.Y. 1972. The ecology of green leafhoppers (mimeographed). IRRI, Saturday Seminar. p. 12.

GHOSH, A. 1985. Rice tungro virus disease situation in Andhra Pradesh. Paper presented at Annual Rice Workshop. Directorate of Rice Research, Hyderabad, India.

HARUKAWA, C. 1951. Effect of environmental temperature upon rice leafhoppers. *Botyu. Kagku.* **16**: 1-11.

HASHIOKA, Y. 1965. Effect of environmental factors on development of causal fungus, infection, development and epidemiology in rice blast disease. In: *The rice blast disease.* Johns Hopkins Press, Baltimore, Maryland, p. 153-161.

HIDAKA, T. and VUNGSILBUTR, P. 1971. Study of environmental factors affecting seasonal fluctuation of population density of the rice gall midge in Thailand Int. Rice Res. Conf. p. 16. (Mimeographed).

ISRAEL, P. and VEDAMOORTHY, G. 1963. Influence of fertility on incidence of the yellow stem borer of rice, *Schoenobius incertulas. Oryza.* **1**: 112-118.

ISRAEL, P., RAO, Y.S. and RAO, P.S.P. 1967. Influence of potash on gall fly incidence in rice. *Oryza.* **5**: .

JADHAV, V.T., MAYEE, C.D., BIDGIRE, M. and SHIRSIKAR, S.P. 1986. Effectiveness of ediphenphos in reducing blast of upland paddy. *Pesticides.* **20**: 24-26.

KALODE, M.B. 1971. Biochemical basis of resistance to brown planthopper and green leafhopper in some rice varieties. IRRI Saturday Seminar. (Mimeographed). p. 1-39.

KALODE, M.B. 1983. Losses due to leafhoppers and planthoppers in rice. Natl. Seminar on Crop losses due to insect pests. *Ind. J. Entomol.* (Special Issue). p. 164-176.

KALODE, M.B. 1984. Pest monitoring in rice. Natl. Seminar on *Concepts in Integrated Pest and Disease Management*, Tamil Nadu Agric. Univ., Coimbatore, September 5-6. (Mimeographed).

KALODE, M.B. 1974. Changes in relative pest status of rice insects as influenced by cultural, biological and genetic factors. Int. Rice Res. Conf. (Mimeographed). p. 1-28.

KIM, C.K. and KIM, C.H. 1991. Predicting rice blast outbreaks in Korea. In: *Rice blast modeling and forecasting*. International Rice Research Institute, Philippines. p. 53-68.

KIYOSAWA, S. 1972. Mathematical studies on the curve of disease increase—A technique for forecasting epidemic development. *Ann. Phytopathol. Soc.* Japan. **38**: 30-39.

KOHLI, C.K. 1966. Pathogenicity and host range studies on the paddy sheath blight pathogen. *J. Res. Pantnagar Agric. Univ.*, India. **3**: 37-40.

KRISHNAN, P., MURALIDHAR, T.T., MANIBHUSHANRAO, K. and CHENCHUREDDY, B. 1992. Epidemiological studies of rice blast disease. 1. Spore releasing capacity of various types of blast lesions and spore release pattern against weather parameters. *Int. J. Trop. Pl. Dis.* **10**: 153-165.

MANIBHUSHANRAO, K. 1994. Rice blast disease. Daya Publishing House, New Delhi. p. 180.

MANIBHUSHANRAO, K. and KRISHNAN, P. 1991. Epidemiology of blast (EPIBLA): a simulation model and forecasting system for tropical rice in India. In: Rice blast modeling and forecasting. International Rice Research Institute, Philippines. p. 31-38.

MAYEE, C.D. 1995. Pest and disease: Weather influence. In: *Agri droughts: database management* Vol. 2. Second SERC School on agrometeorology, Indian Institute of Meteorology and Indian Institute of Tropical Meteorology, Pune. p. 186-194.

MURALIDHARAN, K. 1988. Emerging disease problems in rice—False smut. In: Proc. Rice Pathology Group Discussions. Directorate of Rice Research, Hyderabad, India. (Mimeographed).

MURALIDHARAN, K. and VENKATARAO, G. 1979a. Certain observations on blast disease outbreak and control. In: Golden Jubilee Souvenir, Agric. Res. Stn., Maruteru, Andhra Pradesh, India. pp. 89-91.

MURALIDHARAN, K. and VENKATARAO, G. 1979b. Bacterial blight (*Xanthomonas campestris* pv. *oryzae*) on rice in Nellore district, Andhra Pradesh. Indian Phytopath. **32**: 483-485.

MURALIDHARAN, K. and VENKATARAO, G. 1980a. A simple method of forecasting outbreak of rice blast. *Indian Phytopath.* **33**: 560-564.

MURALIDHARAN, K. and VENKATARAO, G. 1980b. Outbreak of sheath rot (*Acrocylindrium oryzae*) on rice at Nellore. *Int. Rice. Res. Newsl.* **5**: 7-8.

MURALIDHARAN, K. and VENKATARAO, G. 1987. Forecasting outbreak of rice blast disease. In: S. Kannaiyan, ed, Advances in Rice Pathology, Tamil Nadu Agric. Univ., Coimbatore, India. p. 111-137.

MURALIDHARAN, K., REDDY, A.A., SANDHYA, P.N., REDDY, P.C., MANGAL SAIN, ANURADHA, P.R. and REDDY, A.P.K. 1991. Farmers' constraints in the adoption of recommended rice blast control measures. Oryza. 18: 229-231.

MURALIDHARAN, K., REDDY, C.S., SATYANARAYANA, K., GHOSH, A. and VENKATARAMAN, S. 1990. *Ustilaginoidea virens* in rice. Extended Summary, Proc. Int. Symp. on Rice Res.: New Frontiers. 1990, Directorate of Rice Research, Hyderabad, India. p. 232-233.

MURALIDHARAN, K., SHINDE, J.E., PASALU, I.C., VENKATARAMAN, S., PRASAD, A.S.R., KALODE, M.B., RAO, A.V. and REDDY, A.P.K. 1988. Rice in Eastern India—A reality. Oryza. **25**: 213-245.

MURALIDHARAN, K., VENKATARAMAN, S., GHOSH, A., SATYANARAYANA, K., and REDDY, C.S. 1988. Increasing rice production in India through disease surveillance. In: Natl. Workshop Pest and Disease Surveillance for IPM. September 16-17, 1988. Tamil Nadu Agric. Univ., Coimbatore, India.

NAGARAJAN, S., MURALIDHARAN, K. 1995. Dynamics of plant diseases. Allied Publishers. Hyderabad, Bombay, New Delhi.

NARAIN, A. 1988. Impact of brown spot and bunt diseases on rice productivity. In: Perspectives in Mycology and Plant Pathology. MPH, New Delhi. p. 224-243.

ONO, K. 1965. Principles, methods and organisation of blast disease forecasting. In: *The rice blast disease*. Johns Hopkins Press, Baltimore, Maryland. p. 173-194.

OU, S.H. 1985. Rice diseases. Commonwealth Mycological Institute, Kew, Surrey, England. p. 380.

PADMANABHAN, S.Y. 1965. Breeding for blast resistance in India. In: *The Rice Blast Disease*, Baltimore, Maryland: John Hopkins Press. p. 343-359.

PADMANABHAN, S.Y. 1973. The Great Bengal Famine. *Annu. Rev. Phytopath.* **11**: 11-26.

PADMANABHAN, S.Y. 1974. Control of rice diseases in India. *Indian Phytopath.* **27**: 1-28.

PADMANABHAN, S.Y., CHAKRABARTY, N.K. and ROW, K.V.S.R.K. 1971. Forecasting and control of rice diseases. Indian Acad. Sci. **37**B: 423-429.

PADWICK, G.W. 1950. Manual of rice diseases. Commonwealth Mycological Institute, Kew, England.

PARCER, C.S. and CHAHAL, D.S. 1963. Sheath blight rice caused by *Rhizoctonia solani*—a new record in India. *Curr. Sci.* **32**: 328-329.

PATHAK, M.D. 1968. Ecology of common insect pests of rice. *Annu. Rev. Ent.* **13**: 257-294.

PILLAI, K.G., KALODE, M.B. and RAO, A.V. 1979. Effect of nitrogen levels, plant spacing and row orientation on the incidence of the brown plant hopper of rice. *Ind. J. Agric. Sci.* **49**: 125-129.

PRASAD, A.S.R., PRASAD, G.S.V. and MURALIDHARAN, K. 1995. Fragile ecosystem: Fragile for rice? In: *Fragile Lives in Fragile Ecosystems*, Int. Rice Res. Inst., Philippines. p. 369-393.

RAMAKRISHNAN, T.S. 1971. Diseases of Rice, Indian Council of Agricultural Research, New Delhi. p. 145.

RAO, A.V. and KANNAIYAN, S. 1973. An easy method of screening varieties for resistance to sheath blight diseases. *Indian J. Mycol. and Plant Pathol.* **3**: 106-107.

RAO, P.S.P., ISRAEL, P. and RAO, Y.S. 1971a. Epidemiology and control of the rice hispa, *Dicladispa armigera* Oliver. *Oryza* **8** (Suppl.): 345-359.

RAO, P.S.P., RAO, Y.S. and ISRAEL, P. 1971b. Factors favouring incidence of rice pests and methods of forecasting outbreaks: Gall midge and stem borers. *Oryza* **8**: 337-344.

REDDY, A.P.K. 1983. Bacterial leaf blight of rice and its control. Paper presented at Annual Rice Workshop, Hyderabad, India. April 9-12, 1983.

REDDY, C.S. and GHOSH, A. 1985. Sheath rot incidence and yield loss in rice due to joint infection of rice tungro virus and sheath rot fungus. *Indian Phytopath.* **38:** 165-167.

REDDY, D.B. 1964. The rice gall midge *Pachydiplosis oryzae* (Wood-Mason). In: The Major Insect Pests of Rice Plant. John Hopkins Press, Baltimore. p. 457-491.

SADASIVAN, T.S., SURYANARAYANAN and RAMAKRISHNAN, L. 1965. Influence of temperature on blast disease. In: *The Rice Blast Disease*, Baltimore, Maryland: John Hopkins Press. pp. 163-171.

SAHA, N.N. and SAHARIA, D. 1970. Effect of dates of transplanting and levels of nitrogen on the incidence of stem borer, *Tryporyza incertulas* (Walker) in paddy in Assam. *Ind. J. Entomol.* **32**: 225-229.

SAKSENA, H.K. and CHAUBEY, R.D. 1972. Banded blight disease of paddy, paper presented at AICRIP Annual Rice Workshop held at Kanpur, Directorate of Rice Research, Hyderabad, India.

SHANMUGAM, N. and JEYARAJAN, R. 1988. Rice diseases in Tamil Nadu. Paper presented at Rice Pathologists Group Meeting, Jan. 16-17, 1988. Directorate of Rice Research, Hyderabad, India. (Mimeographed).

SHARMA, S.K. and MEHROTRA, R.S. 1988. Stem rot disease of rice-problem and prospects. In: Perspectives in Mycology and Plant Pathology. MDH, New Delhi. pp. 187-211.

SIDDIQ, E.A., MURALIDHARAN, K., GHOSH, A., VENKATARAMAN, S., KRISHNAIAH, N.V. and RAO, J.R.K. eds. 1994. Management of Rice Tungro Virus. 2nd edition. Directorate of Rice Research, Hyderabad, India.

SINGH, R.A. and DUBE, K.S. 1978. Assessment of loss in some rice cultivars due to false smut. *Indian Phytopath.* **31**, 186-188.

SRINIVASAN, M.C., THIRMALACHAR, M.J. and PATEL, M.K. 1959. Bacterial blight disease of rice. *Curr. Sci.* **28**: 469-470.

SRIVASTAVA, D.N. and RAO, Y.P. 1963a. Epidemic of bacterial blight disease of rice in north India. *Indian Phytopath.* **16:** 393-394.

SRIVASTAVA, D.N. and RAO, Y.P. 1963b. Seed transmission and epidemiology of the bacterial leaf blight disease of rice in north India. *Indian Phytopath.* **14**: 77-78.

SUBBAIAH, J. 1986. Major disease problems in rice and their management in Andhra Pradesh. In: Diamond Jubilee Souvenir, Agric. Res. Station, Maruteru, Andhra Pradesh, India. p. 110-117.

SUZUKI. H. 1969. Diurnal periodicity in spore discharge of rice blast fungus. *Rev. Plant Prot. Res.* **2**: 64-65.

SUZUKI, H. 1975. Meteorological factors in the epidemiology of rice blast. *Anu. Rev. Phytopath.* **13**: 239-253.

TENG. P.S., KLEIN-GEBBINCK, H.W. and PINNSCHMIDT, H. 1991. An analysis of the blast pathosystem to guide modeling and forecasting. In: *The Rice Blast Disease.* Johns Hopkins Press, Baltimore, Maryland. p. 1-30.

VANDERPLANK, J.E. 1963. Plant Diseases: Epidemics and Control. Academic Press, New York. p. 349.

VARADARAJAN, G. and NAGARAJA. RAO, P.R. 1965. Influence of nitrogen and phosphoric acid on the incidence of the rice stem borer, *Tryporyza incertulas.* FAO Rice Newsl. 13.

VENKATARAMAN, S. and MURALIDHARAN, K. 1988. Sheath rot disease of rice. In: Proc. Natl. Sem. *Basic Res. for Crop Disease Management.* Tamil Nadu Agric. Univ., Aduthurai, Tamil Nadu.

VENKATARAMAN, S., GHOSH, A. and MAHAJAN, R.K. 1987. Synergistic effect of rice tungro virus and *Sarocladium oryzae* in sheath rot disease of rice. *Int. J. Pl. Dis.* **5**: 141-145.

VENKATARAMAN. S., MURALIDHARAN, K., GHOSH, A., SATYANARAYANA, K. and REDDY, C.S. 1990. Lowland rice diseases and their management. In: *Rice in Wetland Ecosystem,* Kerala Agric. Univ., India. pp. 181-188.

VENKATARAO, G. and MURALIDHARAN, K. 1977a. Outbreak of rice hospa in Nellore district, Andhra Pradesh. *Int. Rice Res. Newsl.* **2**(5): 18-19.

VENKATARAO, G. and MURALIDHARAN. K. 1977b. Outbreak of malayan black bug in nellore district, Andhra Pradesh. *Int. Rice Res. Newsl.* **2**(6): 16.

VENKATARAO, G. and MURALIDHARAN, K. 1982a. Effect of meteorological conditions on the incidence and progress of blast disease of rice. *Z. Pflakrant. Pflalshutz.* **89**: 219-223.

VENKATARAO, G. and MURALIDHARAN, K. 1982b. Plant population density and severity of rice blast disease. *Indian Phytopath.* **35**: 318-320.

VENKATARAO, G., RAJAN, C.P.D. and MURALIDHARAN, K. 1988. Current status of blast and sheath blight diseases of rice in Andhra Pradesh. In: Proc. Rice Pathology Group Discussions. Directorate of Rice Research, Hyderabad. (Mimeographed).

YAMANAKA, I. 1965. Forecasting technique in warm southern Japan based on weather conditions. *Ann. Phytopathol. Soc.*, Japan. **31**: 278-282.

Rice—in a Variable Climate, 1999, pp. 197-215
Eds. Y.P. Abrol and Sulochana Gadgil

11. Methane Emission

Methane gas occurs naturally in the earth's atmosphere. Its concentration has been increasing during the last hundred years. This is because of anthropogenic activity and is strongly correlated with the increase in human population during this period. At present, the atmospheric concentration is more than 1720 ppbv and is highest ever during the last 16,000 years.

Methane has a strong absorption band near 7.7 micron in the infrared radiation emitted by the earth's surface. This is known as 'green house effect' as the gas is called 'green house gas'. On a molecule to molecule basis, methane is twenty times more effective than carbon dioxide in trapping infrared radiation. Hence, an increase in the concentration of methane can contribute to warming of the earth's surface. Further, methane being a highly reactive gas, can change the concentration of water vapour, can also lead to formation of stratospheric clouds that can cause the depletion of ozone in the polar regions.

11.1. Sources and Sinks

There is a large uncertainty in the estimate of sources and sinks of methane. Since methane is produced by biological processes under anaerobic conditions, any factor affecting the physical, chemical or biological characteristics of the soil will influence the methane emission rates. The estimates of sources and sinks of methane have been provided by Prather *et al.* (1995). About 70 % of methane emission can be attributed to human activity while 30 % of the emission is from natural sources such as wetlands. The natural and anthropogenic sources are estimated to be in the range of 110-210 (Tg) and 300-450 Tg per year, respectively. Among the natural sources of methane are wetlands, termites and oceans. The anthropogenic sources are either related to fossil fuels or biogenic carbon. The former includes natural gas, coal mines, petroleum industry and coal combustion. The latter sources are from rice paddies, biomass burning, land fills, animal wastes and domestic sewage (see Table 1). In this paper, the emission of methane from rice paddies is discussed.

Table 1. Estimated sources and sinks of methane (Tg/yr)

Sources	
Natural	
Wetlands	115
Termites	20
Ocean	10
Freshwater	5
CH_4 Hydrate	5
Anthropogenic	
Coal mining, natural gas & petroleum industry	100
Rice paddies	60
Enteric fermentation	80
Animal Wastes	25
Domestic sewage	25
Landfills	30
Biomass burning	40
Sinks	
Atmospheric (tropospheric + stratospheric) removal	470
Removal by soils	30
Atmospheric increase	32

Source: Climate Change 1992 The supplementary report to the IPCC Scientific Assessment, pp. 35.

11.2. Rice Paddies

Flooded rice paddies are considered as one of the culprits in increasing the atmospheric concentration of methane. Submerged conditions, which are predominantly anaerobic, as encountered throughout the growing period of rice plant or at some growth stages of the plant, promote the production of methane, the end product of anaerobic decomposition of organic matter (native or added). Continuous flooded conditions in rainfed low lands, as prevalent in Eastern India, can provide various favourable conditions for production of methane. According to estimates, the atmospheric concentrations of methane is increasing by 0.9 % per year. This increase has been attributed to intensive rice cultures, at least in the past, to meet the demands of the ever increasing population especially in South and South East Asia. Rice cultivation will continue to increase to meet the population demand for food and with it a greater threat of increasing methane emission from the sources.

Methane is produced by the anthropogenic bacteria. It may be released into the atmosphere by three different pathways as shown in figure 1. In the atmosphere, methane is oxidized to carbon dioxide by hydroxyl radicals (Crutzen 1991). In soils and in flooded rice paddies, methane is oxidized by micro-organisms, called methanotrops,

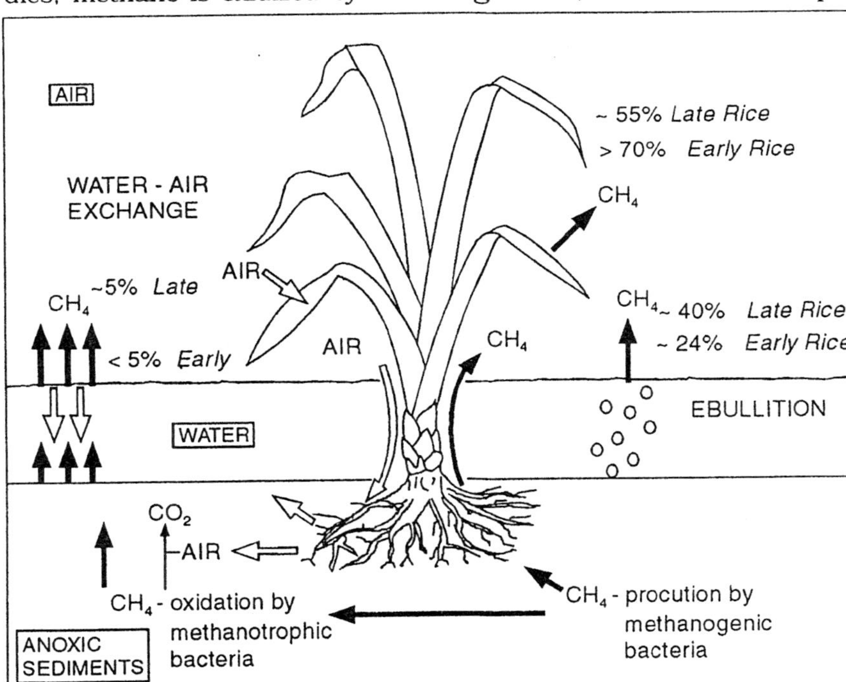

Fig. 1: Methane emission for rice paddy fields (Minami; 1998)

which use methane as an energy source. Holzapfel-Pschorn *et al.* (1985) compared methane production rates estimated from laboratory incubation experiments with emission rates in rice fields and concluded that 80 % of the methane was oxidized in the ecosystem. About 60 to 90 % of the methane produced during a cropping season is oxidized when diffusing into the rice rhizosphere or into the aerobic flood water-soil interface. Schutz *et al.* (1989) observed seasonal variation in oxidation rates. They found that during the early phase of the vegetation period, more than 50 % of the methane was oxidized whereas in the late stage of the vegetation period, 90 % of the methane was oxidized. In addition, methanotrops can oxidize methane to carbon dioxide. Nesbit and Breitenbeck (1992) have shown that the application of nitrogenous fertilizers can decrease the methane oxidation rates in the soil. Large variations in the estimate of methane emission from rice paddies have been reported (Table 2). This is very likely due to rice being grown in a variety of climate, soil and hydrological conditions. It is produced in both tropical and mid-latitude regions and in altitudes ranging from sea level to 2500 meters above the sea level and in temperatures ranging from 4°C to 40°C. A number of studies have shown that methane production is a strong function of temperature and rainfall.

Table 2. Estimates of release rates of methane from rice paddies#

Reference, year	Release of CH_4 (Tg/yr)
Koyama, 1964	190
Ehhalt & Schmidt, 1978	280
Cicerone & Shetter, 1981	59
Khalil & Rasmussen, 1983	95
Seiler *et al*, 1984	35-59
Black, 1984	142-190
Crutzen, 1985	120-200
Holzapfel-Pschorn & Seiler, 1986	70-170
Cicerone & Oremland, 1988	60-170
Schutz *et al*, 1989	50-150
Aselman & Crutzen, 1989	60-140
Schutz & Seiler, 1989	70-170
Wang *et al*, 1990	60-120
Neue *et al*, 1990	25-60
Bouwman, 1990	53-114
Yagi & Minami, 1990	22-73
IPCC, 1990	25-170
IPCC, 1992	20-150
Minami and Yagi, 1993	12-113
Parashar *et al*, 1994*	20

* Parashar, D.C., Gupta, Prabhat K. & Mitra, A.P., Proceedings of TNO/EURASAP Workshop on reliability of VOC emission data base held at Delft, The Netherlands from 9-10 June 1993.

Source: IPCC Greenhouse Gas Inventory Reference Manual Vol. 3, 1995 pp. 4.50.

11.3. Agronomic Studies

11.3.1. Flooded and Non-flooded Conditions

Adhya *et al.* (1994) reported about 20 times higher methane flux in plots planted with rice than those not planted with rice. Ten times more pronounced methane emission under continuously flooded conditions were estimated (Mishra *et al.* 1997). Intermittently flooded regions emitted distinctly less methane than the continuously flooded system. While a significantly negative correlation wss found between methane emission under different water regimes and rhizosphere redox potential, extractable Fe^{++}, readily mineralizable carbon and root biomass presented a significant positive correlation with cumulative methane emission.

11.3.2. Effect of Organic Amendments

Application of farm yard manure (FYM) on rice @ 12.5 t/ha which resulted in an increased emission initially, subsided subsequently. However, application of green manure @ 6.25 t/ha increased the emission at all stages (Table 3 and 4). Combining green manure with low (60:30:30 kg) or high (120:60:60 kg) NPK increased the methane emission than that with green manure alone.

Table 3. Effect of organic amendmentts on methane emission (ADT 36 Kuruvai 94)

Treatment	Methane flux (mg/m^2/h)		
	50 DAT	60 DAT	70 DAT
FYM @ 12.5 t/ha	24.4	30.7	22.7
FYM @ 12.5 t/ha + Low NPK	11.8	15.7	10.8
FYM @ 12.5 t/ha + High NPK	42.2	56.9	14.3
GM @ 6.25 t/ha	41.0	47.1	16.7
GM @ 6.25 t/ha + Low NPK	52.6	65.3	10.3
GM @ 6.25 t/ha + High NPK	66.1	60.6	9.6
NPK 60:30:30 kg/ha (Low)	7.0	24.2	17.5
NPK 125:50:50 kg/ha (High)	27.2	16.1	8.2
Control (no organic or inorganic)	29.5	21.2	26.6

(Kuruvai: 1st crop after fallow for 4 months)
FYM : Form Yard Manure : GM : Green manure
NPK : Nitrogen, Phosphorus, Potassium
DAT : Days After Transplantation

Table 4. Effect of organic amendments on methane emission (ADT 36 paddy Summer '94)

Treatment	Methane flux ($mg/m^2/h$)				Organic 'C'	
	30 DAT	40 DAT	70 DAT	80 DAT	I	F
FYM @ 12.5 t/ha	20.4	14.4	1.0	2.1	1.1	0.9
FYM @ 12.5 t/ha + Low NPK	51.8	15.0	12.1	9.3	1.0	0.8
FYM @ 12.5 t/ha + High NPK	37.3	24.3	2.7	5.6	1.1	1.1
GM @ 6.25 t/ha	43.0	26.7	22.7	25.9	1.1	1.1
GM @ 6.25 t/ha + Low NPK	49.9	49.3	72.9	8.9	1.0	0.9
GM @ 6.25 t/ha + High NPK	58.9	39.3	10.3	43.4	1.1	0.8
NPK 60:30 kg/ha (low)	72.5	17.5	5.2	6.9	0.9	0.7
NPK 125:50:50 kg/ha (High)	34.5	18.2	5.5	6.9	0.9	0.7
Control no organic or inorganic	34.9	26.6	26.6	5.2	0.9	0.7

Measured at 9.00 - 10.00 AM; I: Initial; F: Final

FYM : Farm Yard Manure; GM: Green Manure; NPK: Nitrogen, Phosphates and Potassium; DAT: Days After Transplantation

11.3.3. Effect of Gypsum

Sulphates are inhibitors of methanogenesis. Since gypsum is used as a soil conditioner and ammonium sulphate is used as a nitrogen source, the influence of these inorganics on methane emission was studied. The methane flux was comparatively less in plots where no inorganic fertilizer or gypsum was applied. Gypsum application @ 500 kg/ha in two equal split doses, one basal and another at 25 DAT, along with the second top dressing of nitrogen, is generally recommended for rice cultivation since it confers resistance to plants against diseases and also minimizes the nitrogen loss (Table 5). Application of NPK @ 125:50:50 kg/ha with N as urea increased the methane flux while N as ammonium sulphate depressed it beyond 45 DAT. Combining gypsum with these nitrogenous fertilizers decreased the methane flux in general.

Table 5. Effect of gypsum forms of nitrogen on methane emission in rice

Treatments	Eh of soil (mV)	Methane flux ($mg/m^2/ha$)				Grain * yield
		20 DAT	45 DAT	70 DAT	80 DAT	
Control	- 162	12.0	27.6	23.8	29.3	4800
NPK 125:50:50 kg/ha N as urea	- 158	9.0	40.4	11.5	56.3	5100
NPK 125:50:50 kg/ha N as ammonium sulphate	- 178	18.5	32.4	12.5	01.9	5100
Gypsum @ 500 kg/ha	- 134	21.6	38.0	11.0	18.8	5400
NPK 125:50:50 kg/ha N as urea + gypsum 500 kg/ha	- 139	54.4	30.6	21.8	38.9	5900
NPK 125:50:50 kg/ha N as ammonium sulphate + gypsum 500 kg/ha	- 139	40.1	25.9	25.5	17.3	5800

11.3.4. Plant Population

In a short duration Kuruvai crop, a plant population of 66 hills/m^2 is advocated. In yield maximization trials, higher plant densities with a closer spacing is maintained. Therefore, the effect of plant densities (closer, medium and longer spacing) on methane flux (at recommended NPK of 125:50:50 kg/ha) was studied. It was observed that increasing plant densities (closer spacing) enhanced the methane flux (Table 6). At 45 and 70 DAT, the methane flux was higher in 100 hills/m^2 followed by 66 and 50 hills/m^2. It is likely that the roots of plants concentrated over the area in the closer spacing foraged methane on a large scale compared to plants planted with a wider spacing. In the farmer's field, usually contract planting is done and the labourers engaged will not plant very closely and there is difficulty in achieving the advocated 66 hills/m^2. Therefore, the emission in the farmer's field might be lesser. In parts of Tamil Nadu, where 'Ela' type of rice cultivation is in vogue, a plant density of 50/m^2 or less is adopted. Under such plant densities, the emission might be still smaller.

Table 6. Comparibility of methane concentration in calibration standard and paddy field air samples.

Calibration Standard: Secondary Standard of N.P.L.

*Institute/Laboratory	Methane Concn. (oomv)
NPL	1.94
DAP - CSIRO	1.902
NIAES	1.97
Calibration Standard: Standard of Max Planck Instt.. Germany	
MPI	1.795
NPL	1.79

Air samples collected jointly by NPL and NIAES on 21.11.91 from IARI paddy fields using NPL cnamoer technique at 0.15, 30 minute intervals

	Methane Concen. (oomv)			
	0 min.	15 min.	30 min.	
NPL	1.74	2.02	2.25	
*NIAES	1.791	1.967	2.085	Site–I
NPL	2.25	2.00	1.89	
*NIAES	2.231	1.854	1.948	Site–II

*Institutes/Laboratory:

1. NPL (National Physical Laboratory, New Delhi - 12)
2. DAP - CSIRO (Division of Atmospheric Physics, CSIRO, Australia)
3. NIAES (National Institute of Agro Environmental Sciences, Tsukuba, Japan)
4. MPI (Max Planck Institute, Mainz, Germany)

11.3.5. Rice Stubble, Harvested Fields

Paddy field soil without any crop or water stagnation also emits methane. But this is less than that of transplanted. rice. Fields with summer crack emit a negligible quantity of methane. It is interesting to note that rice stubble after harvest also emit methane. However, this is lesser than that by the whole plants. The emission by rice stubble was comparatively less in summer rice than in Kuruvai. The field after ploughing the stubble recorded a lower methane emission than rice fields with stubble. Therefore, it is advisable to plough the field immediately after harvest in order to reduce methane emission.

11.3.6. Farmer's Field

Emissions recorded in the farmer's field around the research station were relatively less than in the research station itself. Perhaps the farmers apply a lesser dose of fertilizer than that recommended. Among the two varieties viz. ADT 36 and TKM 9 cultivated by the farmers, emission was less in TKM 9. These varieties are comparatively robust and sturdy. It is generally accepted that such varieties transport more oxygen to the roots.

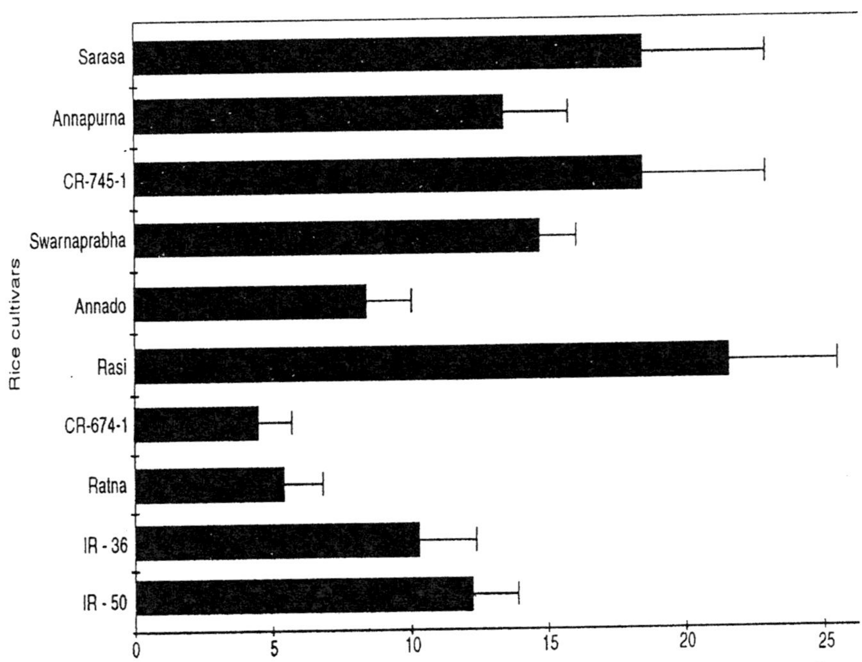

Fig. 2: Cumulative methane efflux from ten rice cultivars planted under uniform field conditions. (The bars indicate standard deviation)

11.4. Cultivar Differences

Wide variations in cumulative ammonium flux amongst rice cultivars has been reported. Methane emission from eight rice cultivars grown under similar conditions differed by as much as one order of magnitude (Parashar *et al.* 1991). Satpathy *et al.* (1998) reported the emission to range from 4.61 gm^2 20.25 g/m^2 (Fig. 2). The cultivars could be classified into three groups on the basis of their methane flux potential. A ranking is also possible based on peak emission occurring at vegetative, reproductive or at both the growth stages. Of the several variables studied in these cultivars (rhizosphere redox potential, above and underground biomass, grain and straw yield, duration of the crop, per cent area occupied by the air space and root oxidase activity), only oxidase activity of the root tip exhibited a significant negative correlation with methane efflux indicating an indirect effect of root oxidation potential on methane efflux.

11.5. Diurnal Pattern

Methane emission showed a distinct diurnal pattern especially at tillering, panicle initiation and maturity stages of field grown rice with maximum in the early afternoon (12.00 to 15.00) followed by a decline to a minimum around midnight. It appears that the redox status in the rhizosphere region and atmosphere, soil and water temperature influence methane emission from rice field by their contrasting effect on methane production and oxidation (Satpathy *et al.* 1997).

11.6. Estimates of Methane Emission from Rice Paddies in India

The emission of methane from Indian rice paddies as estimated by the Environmental Protection Agency (EPA) in USA is more than ten times larger than the results obtained by Parashar *et al.* (1991). EPA estimated that the emission of methane from Indian rice paddies was about 35 Tg per year (*Anonymous*, 1990). Parashar *et al.* (1991) extrapolated the field measurements in some rice growing regions in India and indicated that the methane emission from Indian rice paddies was 3 Tg per year. The methane measurements in India were conducted in a campaign mode in 1991 during the period from June to November in the major rice growing regions in India with diverse agroclimatic environments (Parashar *et al.* 1994). The accuracy of methane measurement was established by calibration and inter-comparison between National Physical Laboratory, New Delhi and other field stations. The absolute field calibration was established by exchanging samples with the Division of Atmospheric Physics, CSIRO, Australia, National Institute of Agro-Environmental Sciences (NIAES), Tsukuba, Japan and National Physical Laboratory, New Delhi. During the campaign, scientists from NIAES visited NPL and a joint air sampling was conducted from paddy fields using the chamber technique. Summary of the results of this comparison is shown in Table 6. The seasonal variation of methane emission from rice paddies in various parts of India is shown in figures 3 to 5. Methane emissions varied

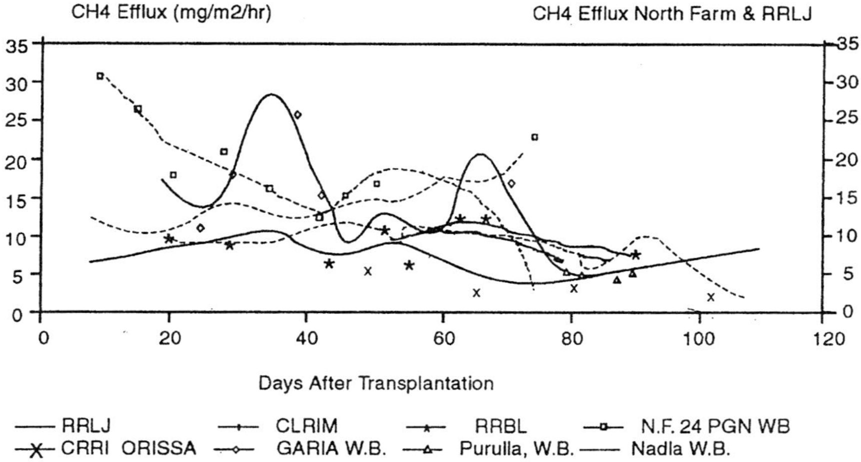

Fig. 3: Seasonal methane emission from 1991 methane campaign stations of rainfed-lowland waterlogged flooded areas. Network stations: RRL-J (-.-), CLIRM (-+-), RRL-B (-*-), North Farm 24 Paraganas W.B. (- -), CRRI (-x-), Garia W.B. (- -), Purulia W.B. (- -), Nadia W.B. (- -).

from around zero to 66 mg per m^2 per hour. The methane emissions from water logged rice paddies was much higher than in the intermittently flooded areas. The large difference in methane emission from

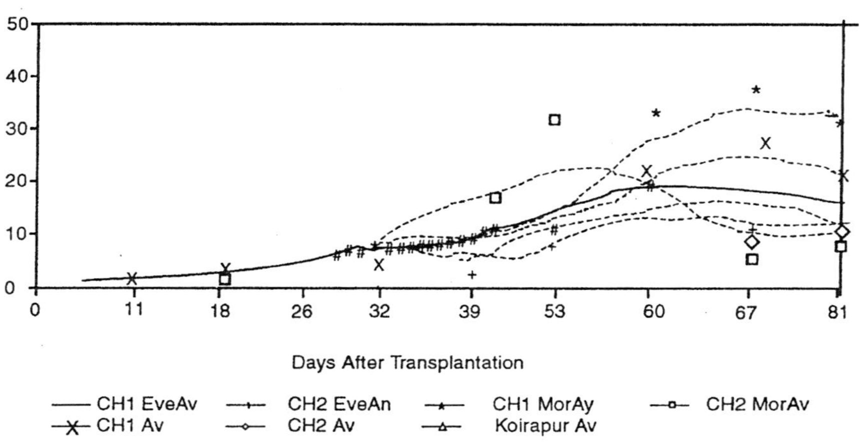

Fig. 4: Seasonal methane emission from 1991 methane campaign station of rainfed lowland deepwater area in Koirapur, Barrackpore, 24 Parganas, North W.B.

different stations was mainly because of differences in rainfall, climate, hydrological conditions and organic carbon content of the soil. The seasonal mean integrated methane flux (SIF) and its range was calculated for each station by taking the daily mean flux and integrating it over the whole cropping season.

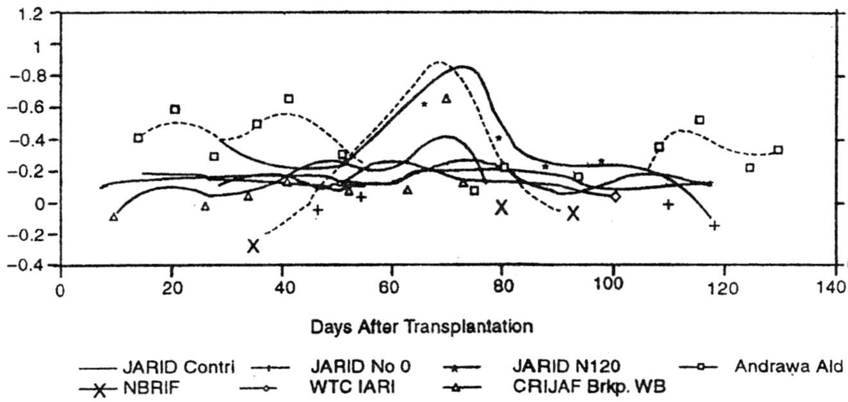

Fig. 5: Seasonal methane emission from 1991 methane campaign stations of intermittently flooded/irrigated areas. Network stations: IARID Control Field (-.-), IARID N60 Field (-+-), IARID N120 Field (-*-), Andawa Allahabad Field (- -), NDAU: Faizabad Field (-x-), IARID WTC Field (- -), CRIJAF Devoke W.B. Field (- -).

The rice growing regions in different parts of India were classified as (see Table 7):

Table 7. Indian paddy cultivation area (mha) distribution

Region/State	Total Area	Continuous Flooded (Irrigated)	Intermilent Flooded (Rainfed)	Deep water	Upland
Eastern Region					
West Bengal	5.61	1.59	2.46	0.68	0.88
Bihar	5.33	1.89	2.24	0.67	0.53
Orissa	4.39	1.44	2.11	0.15	0.69
Assam	2.43	0.53	1.59	0.10	0.22
N.E. States	0.82	0.24	0.22	0.04	0.32
TOTAL AREA	18.58	5.69	8.61	1.69	2.64
Southern Region					
Andhra Pradesh	4.19	2.71	1.25	0.24	-
Tamil Nadu	2.01	1.42	0.59	-	-

Karnataka	1.18	0.71	-	-	0.47
Kerala	0.57	0.24	0.02	-	0.32
Pondicherry	0.03	-	0.03	-	-
Andamans	0.01	-	0.01	-	-
TOTAL AREA	7.99	5.07	1.89	0.24	0.79
Northern Region					
Uttar Pradesh	5.36	1.46	3.17	0.23	0.50
Punjab	1.91	1.91	-	-	-
Haryana	0.62	0.62	-	-	-
J & K	0.28	-	-	-	-
Himachal Pradesh	0.09	-	-	-	0.38
Sikkim	0.02	-	-	-	-
Delhi	0.003	0.003	-	-	-
TOTAL AREA	8.28	3.99	3.17	0.22	0.88
Western Region					
Madhya Pradesh	5.04	0.91	3.38	0.32	1.7
Maharashtra	1.52	0.28	-	-	-
Gujarat	0.60	0.39	0.21	-	-
Dadra N.H.	0.01	-	0.01	-	-
Rajasthan	0.16	0.16	-	-	-
Goa, Daman & Diu	0.05	-	0.05	-	-
TOTAL AREA	7.39	1.74	3.65	0.32	1.67
GRAND TOTAL	42.23	16.50	17.33	2.43	5.97

(a) intermittently flooded (irrigated)
(b) rainfed water logged (flooded)
(c) deep water
(d) upland types

The soil characteristics of different regions are given in Table 8. Methane emission from various regions was estimated using the data from the nearest station with the same type of paddy cultivation technique. Large methane emissions came primarily from rainfed waterlogged paddy fields and not from intermittently flooded and upland areas. The methane emission rates during the growing season were 10 to 50 grams per m^2 in waterlogged and deep water regions and zero to two grams per m^2 in intermittently flooded (irrigated areas). The number of growing days is also important in the calculation of

Table 8. Paddy soil characteristics of 1991 methane campaign sites

Sr. No.	Network station	Type/Texture	pH	Organic carbon %
1.	NPL-IARI	Alluvial Sandy Loam	8.3	0.4–0.8
2.	Allahabad, U.P.			
	a. Andawa	Alluvial Sandy Loam	7.2	0.48
	b. Dehri-on-sone	–do–	5.6	1.56
3.	Faizabad, U.P.	Alluvial	8.2	0.27
4.	CRRI, Cuttack Orissa	Alluvial Sandy-clay	6.2	0.6
5.	RRL-Bhub. CRRI, Cuttack, Fields, Orissa.	–do–	7.2	0.6
6.	Assam			
	a. RRL, Jorhat	Clay Loam	4.7	1.1
	b. RRS, Titabar	Clay Loam	4.6	1.8
7.	Thiruvananthapuram Kerala	Sandy Clay	6.5	1.46
8.	Barrackpore, W. Bengal			
	a. N. Farms	Clay	7.0	0.22
	b. Devoke	Clay loam	6.8	0.32
	c. Koirapore	–do–	6.8	0.32
9.	Kalyani, Nadia Distt.			
	a. Charsharati	Sandy Loam	7.4–8.0	0.64–0.68
	b. Chinsurah Hooghly Distt.	Alluvial	7.2–7.8	0.67–0.86
	c. Iajnagarpara	Sandy Loam	6.5	0.66
	d. Chakasah	–do–	6.2	0.88
10.	Kalna, Burdwan	–	6.3–6.7	0.7–1.0
11.	24-Parganas (N)			
	a. Mahadebpur Haringhata	Alluvial Clay loam	7.0–8.0	0.63–0.68
	b. Kundibagan	Sandy Loam	7.6–7.7	0.39–1.19
12.	Garia Garcha	–	7.8	1.12

total methane flux during the growing season. The number of growing days is usually assumed to be in the range of 120 to 150 days. In India, the number varies between 73 to 130 days (Table 9). Methane emission rates in different regions of India are shown in Table 10.

The rainfed waterlogged and deep water regions occupy 47 % of the total area under paddy cultivation but contribute to 94 % of the total methane emission. The irrigated and upland regions that constitute 50 % of the total area under paddy cultivation but contribute to 94 % of the total methane emission. The irrigated and upland regions that constitute 50 % of the total area under paddy cultivation contribute to 6 % of the total methane emission. The total methane emission from Indian paddy fields is in the range from 2.5 to 6 Tg per year with a mean around 4 Tg per year. Parashar *et al.* (1996) re-evaluated the methane emissions from Indian paddy fields using the categorization used by the International Rice Research Institute (IRRI) and found that the total methane emissions from the Indian paddy fields was in the range of 5 to 6 Tg per year.

Table 9. Paddy cropping and measurement periods at 1991 methane campaign network stations

Sr. No.	Station/Sites	Cropping period (days)	Measurement Dates	Period Days
1.	Jorhat	106	20.8 to 28.11.91	101
2.	Cuttack/			
	C11d	114	20.8 to 19.10.91	61
	A12bc	110	-do-	61
3.	Bhubaneswar	90	21.8 to 23.10.91	64
4.	Barrackpore/			
	Devoke	110	7.8 to 9.10.91	58
	N. Farm	130	2.8 to 11.10.91	64
	Koirapur	130	5.8 to 21.10.91	78
5.	Kalyani	110	16.8 to 1.10.91	53
6.	Chennai	79*	23.8 to 29.10.91	74
7.	Thiruvananthapuram	73	7.8 to 18.09.91	43
8.	NBRI	110	20.8 to 10.10.91	52
9.	Bihar	105	17.9 to 24.11.91	69
10.	Allahabad	119	21.7 to 10.11.91	113
11.	Delhi/			
	Control	118	18.7 to 1.11.91	106
	N60	118	-Do-	106
	N120	118	-Do-	106
	WTC	118	-Do-	106
12.	Calcutta/			
	Garia	100	8.9 to 24.11.91	78
	Purulia	85	20.9 to 14.11.91	56

*Measurements discontinued due to cyclonic destruction.

Table 10. Methane budget estimates from Indian paddy fields

Areas/Regions/States	Cultivation 6 Area (10 ha)	Integrated Seasonal			Reference	Total Methane Emission		
		Methane Max.	Flux Mean	(g/mm^2) Min.	Stations	Max	(Tg/yr) Mean	Min.
A. RAINFED WATERLOGGED FLOODED AREAS								
III. Eastern Region								
a) West Bengal	2.457	34±10	28	23±6	Garia, Purulia, Kalyani CRIJAF-n. Farm	0.84±0.25	0.69	0.57±0.15
b) Bihar	2.235	26±2	22	19±4	Dehri. Bihar	0.58±0.05	0.49	0.43±0.09
c) Orissa	2.109	14±5	12	10±3	RRLB. CRRIC	0.30±0.11	0.25	0.21±0.06
d) Assam	1.590	60±2	46	35±3	RRLJ	0.95±0.03	0.73	0.56±0.05
e) N.E. States	0.219	60±2	46	35±3	RRLJ	0.13±0.004	0.10	0.08±0.007
II. Southern Region								
a) Andhra Pradesh	1.245							
b) Tamil Nadu	0.593	13±2	11	9±2	CLRIM	0.26±0.04	0.21	0.17±0.04
c) Pondicherry	0.025							
d) Kerala	0.017	13±1	9	5±1	RRLT	0.002±0.0002	0.002	0.001±0.000
e) Andaman & Nicobar	0.012	22±10	17	12±6	W.B. Stations	0.002±0.001	0.002	0.014±0.000
III. Northern Region								
Uttar Pradesh	3.174	26±2	22	19±4	Dehri, Bihar	0.83±0.06	0.70	0.60±0.13

IV.	Western Region								
	a) Madhya Pradesh	3.377	16±8	14	11±4	Orissa, Bihar, & Tamil Nadu	0.54±0.27	0.47	0.37±0.14
	b) Maharashtra								
	c) Gujarat								
	d) Goa, Daman & Diu	0.275	13±1	9	5±1	RRLT	0.036±0.003	0.03	0.014±0.002
	e) Dadra N.H.								
	TOTAL	17.328					4.17±0.82	3.67	3.01±0.67
B.	DEEP WATER AREAS	2.434	24±2	19	14±1	CRIJAF, W.B.	0.58±0.046	0.06	0.342±0.025
C.	INTERMITTENTLY FLOODED-IRRIGATED AREAS	16.497	2±0.6	0.74	0.06±0.01	Delhi, Allahabad, Faizabad, Barrackpore	0.335±0.078	0.122	0.06±0.007
D.	UPLAND AREAS	5.973	Negligible					Negligible	
	GRAND TOTAL	42.232					5.4±1.0	4.3	3.4±0.7

11.7. Conclusions

After water vapour and carbon dioxide, methane is the most important greenhouse gas and hence can influence global climate change. It can play a significant role in controlling the abundance of other greenhouse gases in the atmosphere. The life time of methane is, however, much smaller than that of carbon dioxide and hence its levels can decline quickly if anthropogenic sources are reduced.

A number of mitigation strategies for methane emission are feasible. These primarily involve water management and choice of cultivars (Vamadevan and Jha, 1985; Sass and Fisher, 1998). Also of interest are the observations of Debnath *et al.* (1996) regarding the use of biodigested slurry. It is necessary that the mitigation options keep in view that other desireable characteristics such as resistance to pests, drought and diseases are not affected. In addition, we need not only reduce the concentration of methane but also N_2O emissions which are of great concern.

Indepth studies on basic aspects that influence methane production, consumption and transport particularly in relation to cultivars are required to be conducted. Of particular significance are the findings of Wassmann *et al.* (1993) regarding the cultivar differences in root exudates and root litter. In this context the observations of Ghosh *et al.* (1995) regarding substrate availability vis-à-vis requirement for methane production need attention.

There are large uncertainties in the estimates of contribution of rice cultivation to anthropogenic methane emissions (Parashar *et al.* 1997). It is possibly due to rice being grown in different climate, soil and hydrological conditions. There are problems pertaining to the methodologies being used as well. Extrapolation of results from a few experimental plots to the total rice growing area in the country needs careful examination. It is very likely that emissions from rice paddies may not be an important factor in global warming caused by an increase in the green house gases.

11.8. References

ADHYA T.K., RATH, K.R., GUPTA, P.K., RAO, V.R., DAS, S.N., PARIDA, K.M., PARASHAR, D.C. AND SETUNATHAN, N. 1994. Methane emissions from flooded rice fields under irrigated conditions. *Biol. Fert Soils.* **18**: 243-248.

ADHYA, T.K., PATNAIK, P., SATPATHY, S.N., KUMARSWAMY, S. AND SETHUNATHAN, N. 1997. Influence of phosphorus application on methane emission and production in paddy field soils. *Soil Biol. Biochem.*, pp. 1-5.

ANONYMOUS. 1990. Methane emissions and opportunities for control. US Environmental Protection Agency, EPA/400/9-90/007, Washington DC.

CRUTZEN, P.J. 1991. Methane's sources and sinks. *Nature*. **350**: 380-381.

DEBNATH, G., JAIN, M.C., KUMAR, S., SARKAR, K. AND SINHA, S.K. 1996. Methane emissions from rice fields amended with biogas slurry and farm yard manure. *Climatic Change*. **33**: 97-109.

GHOSH, S., JAIN, M.C. AND SINHA, S.K. 1995. Estimation of global methane production from rice paddies based on substrate requirement. *Curr. Sci.*, **69**: 937-945.

HOLZAPFEL-PSCHORN, A., CONRAD, R. AND SEILER, W. 1985. Production, oxidation and emission of methane in rice paddies. *FEMS Microbiol. Ecol.*, **31**: 343-351.

MINAMI, K. 1998. Methodologies for methane estimation. In: *Global environmental change* (Parashar, D.C., Sharma, C. and Mitra, A.P., eds.), Narosa Publishing House, New Delhi. p. 95-110.

MISHRA, S., RATH, A.K., ADHYA, T.K., RAO, V.R. AND SETHUNATHAN, N. 1997. Effect of continuous and alternate water regimes on methane efflux from rice under green house conditions. *Biol. Fert. Soils*. **24**: 399-405.

NESBIT, S.P. AND BREINTENBECK, G.A. 1992. Factors influencing methane uptake by soils. *Agricultural Ecosystem Environment*. **41**: 39-54.

PARASHAR, D.C., GUPTA, P.K. AND BHATTACHARYA, S. 1997. Recent methane budget estimae from rice paddy fields. *Indian J. Radio and Space Phys*. **26**: 237-243.

PARASHAR, D.C., RAI, J., GUPTA, P.K. AND SINGH, N. 1991. Parameters affecting methane emission from paddy fields. *Indian J. Radio and Space Physics*. **20**: 12-17.

PARASHAR D.C. *et al.* 1994. Methane budget from Indian paddy fields. *Curr. Sci.* **66**: 938-940.

PRATHER, M., DERWENT, R., EHHALT, D., FRASER, P., SANHUEZA, E. AND ZHOU, X. 1995. Other trace gases and atmospheric chemistry. In: Climate change 1994. Radiative forcing of climate change and an evaluation of the IPCC IS92 Emission Scenario, Houghton, J.T., Meira Filho, L.G., Bruce, J., Hoesung Lee, Callander, B.A., Haites, E., Harris, N. and Maskell, K. (eds), Cambridge University Press, Cambridge, U.K.

SASS, R.L. AND FISHER, S.M. 1998. Methane from irrigated rice cultivation. In: *Global environmental change* (Parashar, D.C., Sharma, C. and Mitra, A.P., eds.), Narosa Publishing House, New Delhi. pp. 77-94.

SATPATHY, S.N. MISHRA, S., ADHYA, T.K., RAMAKRISHNAN, B., RAO, V.R. AND SETUNATHAN, N. 1998. Cultivar variance in methane efflux from tropical rice. *Plant and Soil*. **202**: 223-229.

SATPATHY, S.N., RATH, A.K., RAMAKRISHNAN, B., RAO, V.R., ADHYA, T.K. AND SETUNATHAN, N. 1997. Diurnal variation in methane flux at different growth stages of tropical rice. *Plant and Soil*, **197**: 267-271.

SCHUTZ, H., HOLZAPFEL-PSCHORN, A., CORNAD, R., RENENBERG, H. AND SEILER, W. 1989. A three year continuous record on the influence of day, season and fertilizer treatment on methane emission rates from an Italian rice paddy field. *J. Geophys, Res.* **94**: 16405-16416.

VAMADEVAN, V.K. AND JHA, K.P. 1985. *Rice research in India.* ICAR, New Delhi. pp. 263.

WASSMANN, R., PAPEN, H., RENNENBERG, H. 1993. Methane emission from rice paddies and possible mitigation strategies. *Chemosphere.* **26**: 201-217.

Rice—in a Variable Climate, 1999, pp. 217-238
Eds. Y.P. Abrol and Sulochana Gadgil

12. Rice In 2000 A.D.*

M.S. Swaminathan

Introduction

In legend and religion around the world, rice is regarded as a precious gift of God to the earth. Rice has a very wide range of adaptation, growing below sea-level and at 2,000 m altitude. It grows under wide range of latitudes. Further, it grows under drought-prone and flood-prone conditions and in acid soils as well as saline/alkali soils. Rice is also unique in that it is nearly entirely used as human food, unlike other cereals which are also used extensively as feed grains.

The geographical distribution of rice-growing areas in different parts of the world reveals that rice is cultivated from 50° North in Central Czechoslovakia, on the equator, to 35° South in Australia. While the vast majority of rice land is located at low elevations, rice is also found at altitudes over 2000 m. The crop is cultivated under extremely wet conditions with more than 3000 mm rainfall during its growing period, but it is also grown in desert climates with less than 50 mm rainfall during its growing season. In Pakistan, rice is produced in areas with mean monthly temperatures of more than 33° Celsius, while in northern Japan the average season temperature is not much higher than 17°C. Total solar radiation values range from around 40 kilo langleys during the growing season (around 300 MWh cm^{-2} per day) to more than 80 kilo langleys (around 600 MWh cm^{-2} per day) during the growing season.

Since rice is cultivated under such a great variation of climatic conditions it is not possible to define a typical rice climate. For optimum growth and yield, the rice crop needs a combination of several climatic conditions, which seldom occur together in nature.

* A large volume of water evenly distributed throughout the major portion of the growing season but relatively dry during the ripening period.
* Sufficient high temperatures throughout its growing season, but somewhat lower night temperatures during the ripening period.
* Ample solar radiation throughout the growing season.

---

* Reproduced with permission

Present address : M.S. Swaminathan Research Foundation, 3rd Cross Street, Taramani Institutional Area, Chennai 600013

The climate in which rice was domesticated most probably had 5- to 6-month growing season. Since the history of rice is a fascinating one, it may be appropriate to make a brief reference to it.

Pre-Gondwana Origin of *Oryza*

The cultivated rices of the world belong to the genus *Oryza*, which also contains 20 wild species. *O. sativa* L. of Asia and *O glaberrima* Stud. of West Africa are two cultigens. The wild species are scattered in Asia, Africa, Australia, and Central and South America. These species have either 24 to 48 chromosomes. Six different genomes designated A to F have been recognized on the basis of cytogenetic studies.

Based on the pan-tropical and widely dispersed distribution of *Oryza* species, IRRI geneticist T. T. Chang (1976) speculates that the Gondwana supercontinent may have been the original habitat of the genus *Oryza*. When Gondwanaland fractured and became Africa, Antarctica, Australia, Malagasy, South America, and South and Southeast Asia, *Oryza* species drifted into distinct geographical habitats (Figs. 1 and 2).

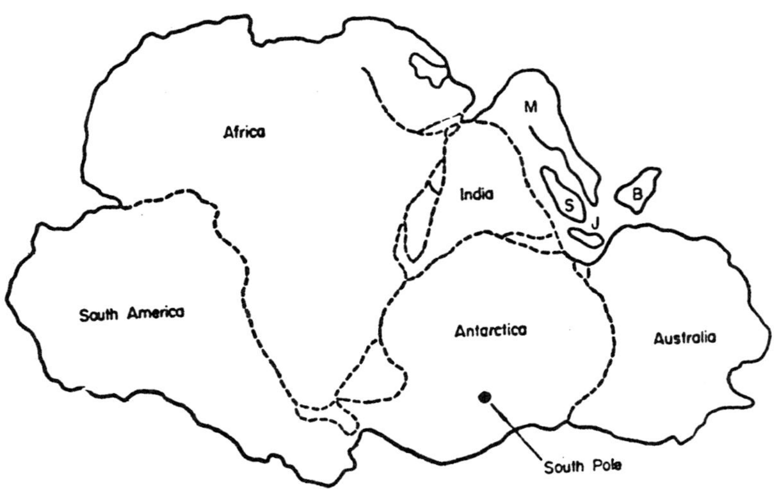

Fig. 1: The Gondwanaland continents before fracture and drift.

Humans domesticated *O. Sativa* in Asia, and *O. Glaberrima* in Africa. Only wild *Oryza* species are found in Australia and South America.

While there is evidence that *O. sativa* was cultivated more than 7,000 years ago, *O. glaberrima* seems to have been domesticated somewhat later in Africa.

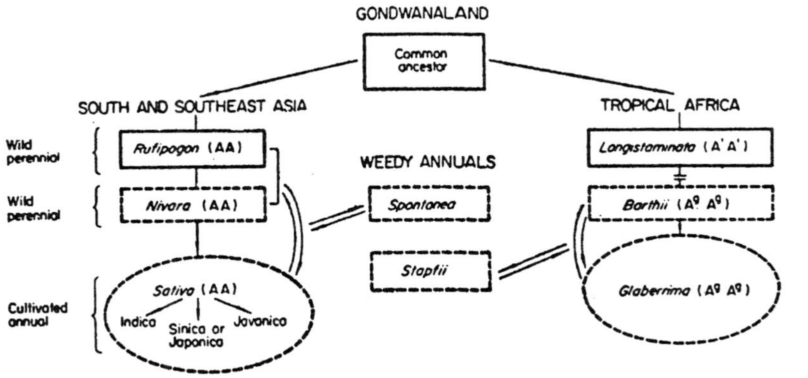

Fig. 2: Evolutionary pathway of the two cultivated species of rice. Taxa boxed by solid lines are wild perennials. Taxa boxed by broken lines are annuals. Arrow with solid lines indicates direct descent. Arrow with broken lines indicates indirect descent. Double arrows indicate introgressive hybridization.

Asian Rices

The Asian cultivated species *O. sativa* differentiated into three eco-geographic races : *indica, japonica* or *sinica*, and *Javanica*. Most rice researchers agree that the area with the greatest *O. sativa* diversity is a belt extending from the Assam-Meghalaya areas of India to the mountain ranges of mainland Southeast Asia and Southwest China. Early-maturing and drought-escaping annual strains probably emerged during the Neothermal period (about B. C. 15,000 to 10,000) along the southern and northern slopes of the Himalayas.

Annual ancestral forms of the Asian cultigen began to appear in the Neothermal age at the periphery of the wild annual progenitors, mainly on the southern borders of Himalaya and to a lesser extent in South and Southwest China. Alternating periods of drought and pronounced temperature variation accelerated the development of the annual forms in Northeast and East India, northern Southeast Asia, and southern China. The early cultigens are *indicas*.

Cultivated varieties matured earlier than their progenitors; therefore, they survived better in drier, fluctuating climates and produced more seeds, which helped them to move farther north than perennial forms.

Increasing aridity on the northern border of Himalaya and its associated mountain ranges in mainland Southeast Asia forced early settlers in China to move east and south, toward more humid areas. As people migrated, plants also dispersed, which greatly accelerated the ecogenetic diversification process.

Inside the present boundaries of China, along a branch of the Brahmaputra River (Hsiangchua River), a temperate-zone race evolved from the primary tropical *indica* race. This race has been widely known as *japonica* since Japanese workers coined the term in 1928.

Tall, large-grained rices migrated to the Indonesian islands from the east coast of South Asia and differentiated to become *javanica* rice.

The oldest remains of cultivated rice, both *indicas*, were found in east China and northern India, and date back 7000 years. The oldest *sinica* (ken) rice found in China is dated 3300 B.C. Grains of what probably was wild rice have been excavated at Non Nok Tha in northern Thailand and date to about 4000 B.C. Rice farmers of Asia are hence referred to as "farmers of 50 centuries".

The later dispersal of Asian cultivars, mainly *indicas*, from India to the Middle East, North Africa, and then Europe probably began around 1000 B.C. Many *javanica* cultivars found their way from Indonesia into Madagascar. West Africa obtained rice varieties from Europe or directly from South Asia. European countries supplied most of the first cultivars grown in South America. The United States obtained rice seeds from Madagascar, South Asia, and East Asia.

Oryza glaberrima in West Africa

Porteres (1956) postulated that the African cultigen *O. glaberrima* originated on the Niger River Delta. The primary centre of diversity for *O. glaberrima* is in the swampy upper Niger River Basin and southwest in two areas near the Guinean coast. Rice was probably cultivated in the Niger basin around 1500 B.C. and was grown in Guinea about 400 years later. In West Africa, *O. glaberrima* is a dominant crop grown in flooded areas of the Niger and Sokoto River basins.

The African cultigen and its annual wild race (*O. barthii*) are less diverse than their Asian counterparts. Although Porteres recognized two subspecies designated *vulgaris* and *humilis* in *O. glaberrima*, Oka and Morishima (1971) indicated that the *barthii-glaberrima* complex could be differentiated into two ecotypes : deep water and upland. African cultivars are inferior to the Asian cultivars in deep water because they have limited internode elongation ability.

The differentiation of *O. sativa* in Asia probably predated that of *O. glaberrima* in West Africa, based on their history of domestication and the extent of intraspecies varietal diversification.

Differentiation of *Indica, Japonica,* and *Javanica* Races

Indica rices were confined to the humid regions of the Asian tropics and subtropics. *Japonicas* were cultivated in subtropical and temperate regions. *Javanicas* occurred in the equatorial region in Indonesia.

In addition to climate adaptation, these races also differ in grain quality characteristics such as amylose content, gelatinization temperature, gel consistency, grain elongation, and aroma. Cooking and eating qualities are largely determined by the properties of starch, which constitutes 90 per cent of milled rice. Differences in grain seem to have developed by selection based on consumer preferences. *Japonica* varieties planted in Taiwan were named *pon-lai* rice, meaning heavenly rice, the term being derived from *horai* in Japanese.

Javanica varieties, also known as *bulu* types, are distinguished from *japonicas* and *indicas* by stiff straw; tall plant stature; low tillering habit; long panicles; long awns; large, bold non-shattering grains; and low photoperiod sensitivity (number of hours of sunlight). *Bulus* are still grown in Bali and Lambok, on the mountain terraces of Banaue in the Philippines, and in the mountains of Taiwan.

The genetic affinities among *indica*, *japonica*, and *javanica* subspecies have been studied by several scientists using as criteira the fertility of F_1-hybrids, chromosome pairing behaviour during meiosis, and F_2-segregation pattern. Early studies by Japanese scientists showed that chromosomal aberrations such as loose pairing, univalents, quadrivalents, inversions, and deletions were found at meiosis in several *japonica-indica* hybrids. Later studies at IRRI revealed that similar chromosomal aberrations develop if *indica* varieties from different areas are crossed, especially the tall *indicas* and the Chinese semidwarf strains.

The available data indicate that both inter-and intra-racial sterility and temporal-geographic isolation may play a significant role in producing the genetic barriers. Moreover, chromosomal aberrations found in partially sterile hybrids cannot account for all F_1-incompatibility, thus indicating that genetic imbalance may also be involved in the expression of sterility.

Renewed interest in ecogenetic differentiation has led to recent studies involving physiological and biochemical analyses. The findings pointed out the inadequacy of relying heavily on gross plant and grain morphology to characterize and classify the numerous rice cultivars. The grain size and shape of rices are now known to be subject to the rice grower's and consumers' preferences, and the predominant grain type of a region can be altered, sometimes in a matter of decades. Other criteria useful to a fuller understanding of the varietal complexes are amylose content of the endosperm, temperature and photoperiod responses, edaphic-hydrological adaptation expressed in root characters, and grain ripening duration.

Grain quality is an attribute that must be defined in relation to a location. For example, in Thailand good quality rice has long, slender, translucent grains that produce a fluffy, tender, cooked product. In Japan, rices with short, broad grains and sticky cooking properties that make eating with chopsticks easy are preferred. In parts of Northwest India and Pakistan, *basmati* varieties with long, slender, scented grains fetch a premium price.

The ecogenetic differentiation process was primarily achieved by natural and human selection in different environments following hybridization, segregation, or mutation. Most differentiation occurred in the broad belt extending from the southern foothills of the Himalayas to Vietnam, and on the northern borders of the mountain ranges in Southwest and South China. Today, in the hills of Yamuna and Kweichow Province (China), *japonica* (keng) rices are grown at elevations above 1,800m, a mixture of *japonica* and *indica* rices at the medium elevations, and *indica* (hsien rices) below 1,000m.

Recent History of Rice Improvement

Until the end of World War II, rice yields in most countries of Asia were stagnating at a level of 1 to 2 t/ha. While this was the situation with respect to *indica* varieties of rice, *japonica* varieties cultivated in Japan, USA, and the Mediterranean region of Europe yielded about 5 t/ha. It was then believed that the *japonica* varieties have the genetic capability for responding to fertilizer application, a trait which *indica* varieties lacked. On the basis of this assumption, an extensive *indica-japonica* hybridization programme was initiated in 1952 with the Central Rice Research Institute at Cuttack being the primary hybridization centre. Because of partial sterility and skewed recombination, very few commercial varieties came out of this programme. Some good varieties developed from *indica*×*japonica* crosses are 'ADT 27' in Tamil Nadu in India and 'Mashuri' in Malaysia.

Following the establishment of the International Rice Research Institute at Los Banos in the Philippines in 1960, it soon became apparent that what is needed in *indica* rices is a new plant type which is capable of utilizing effectively both solar energy, and nutrients and water from the soil. Herdt and Capule (1983) have recently compiled statistics on the adoption and spread of modern rice varieties of rice in the mid-1960s marked the beginning of an accelerated advance in the production and productivity of *indica* rices. Data relating to the growth of rice production in the different regions of the world and India are given in Tables 1 and 2.

As will be seen from Table 1, North and Central America as well as South America registered high growth rates of 5.5 per cent and 3.4 per cent respectively. In contrast, the growth rate during the 70s in Asia and the Pacific regions was 2.4 per cent. In India, modern varieties were planted on 44 per cent of the rice area in 1978. In Jammu & Kashmir, Punjab, Tamil Nadu, Andhra Pradesh, Kerala and Maharashtra more than 60 per cent of the area was planted to MVs in 1978; Assam and Bihar had less then 30 per cent. Other states

Table 1. Growth of rice paddy production, 1971-81

Geographic entity	Production (million tons)		
	Annual average for triennium ending 1971	Annual average for triennium ending 1981	Compound annual growth rate, %
Asia & Pacific	282.52	357.88	2.4
Rest of the World	27.27	38.37	3.5
Africa	7.34	8.41	1.4
North & Central America	5.35	9.11	5.5
South America	9.57	13.37	3.4
Europe	1.82	1.92	0.5
Others	3.19	5.56	5.7
World	309.79	396.25	2.5

Source : FAO 1983

had intermediate adoption rates. Differences in adoption are reflected in yield differences across the states. Rice yields show a sharp steady rise in the northern states of Jammu & Kashmir, Punjab, and Haryana. In the southern states, viz., Tamil Nadu, Karnataka, Andhra Pradesh, and Kerala, yields have risen, but not as sharply as in the north. In eastern India yields have changed little, because eastern India, like Bangladesh, requires better varieties to achieve increased rice production (Fig. 3).

Table 2. Spread of MVs and their comparative yields with other rices India, 1966-78
(India Directorate of Economics and Statistics 1978)

Year	Area (thousand ha)		% area in MVs	Comparative yield* (t/ha)		National average yields
	MVs	All rice		MVs	Others	
1966-67	888	35,598	2.4			1.3
1967-68	1,785	36,437	7.8			1.6
1968-69	2,681	36,966	7.2	3.1	1.6	1.6
1969-70	4,253	37,680	11.2	2.8	1.6	1.6
1970-71	5,454	37,592	14.5	3.1	1.6	1.7
1971-72	7,199	37,758	19.0	3.2	1.4	1.7
1972-73	8,107	37,688	22.0	3.0	1.2	1.6
1973-74	9,718	38,285	25.3	3.0	1.3	1.7
1974-75	10,780	37,888	28.4	1.7	1.0	1.6
1975-76	12,742	39,475	32.2	2.8	1.4	1.9
1976-77	13,731	38,511	35.6	2.5	1.2	1.6
1977-78	15,516	40,283	38.5	2.5	1.4	2.0
1978-79	17,619	40,482	43.5			2.0

* Area, production, and yield data are recorded separately for HYVs and other rices of some states each year. Thus the MV yield data do not include all rice grown in the country.

Source : Herdt and Capule 1983.

An encouraging feature of recent developments in the improvement of rice production is the fact that much of the increase has come from yield improvement. This is particularly important in Southeast Asia where the per capita availability of land is already low, being about 0.21 ha per individual. Data relating to the pathway of production advance in different parts of the world are given in Table 3. The highest national growth rate, of 9.9 per cent, was recorded in Australia. But this is from a very low base level of absolute production. Among the large paddy-producing countries, Democratic People's Republic of Korea had a most significant growth rate 7.2 per cent, followed by Burma 4.6 per cent, Indonesia 4.5 per cent and the Philippines 3.9 per cent. There are a few countries in the region where the progress was marginal or even negative. For instance, Kampuchea had a negative growth of 11.6 per cent, In Fiji, growth rate was also negative and so was in Japan. In Nepal, the production remained practically stagnant.

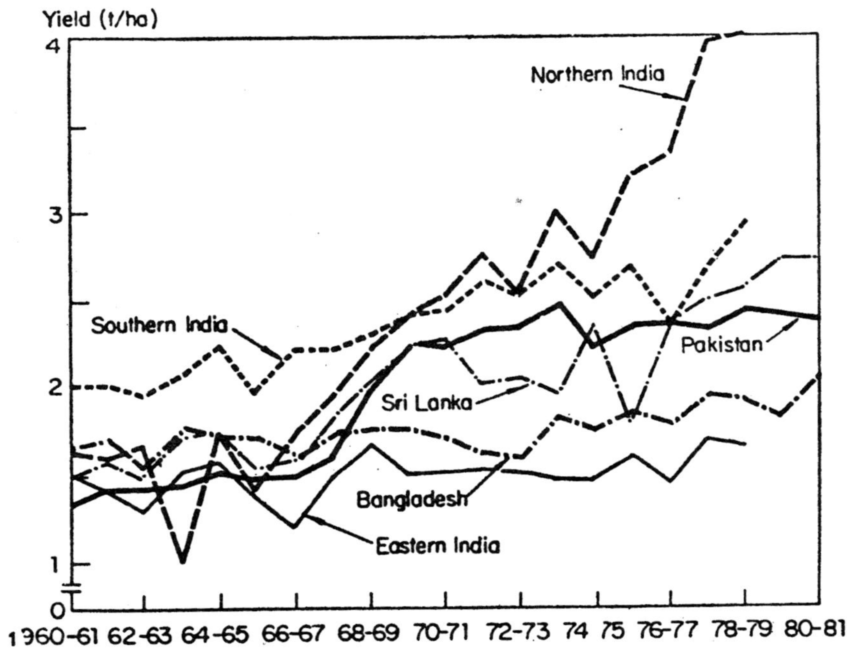

Fig. 3: Rice yield trends in 3 regions in India, Bangladesh, Pakistan, and Sri Lanka, 1960-61 to 1980-81.

The contribution made by increase in yield and area expansion to production growth in rice paddy is indicated in Table 4. It will be seen from the data that while in Burma, Indonesia, Philippines and China rice paddy yields increased by 2.5-4 per cent per annum, there were some countries, such as Kampuchea, Nepal, Sri Lanka and

Table 3. Rice paddy yield and area, 1971-81

Geographic entity	Yield (kg/ha)			Area (million/ha)		
	Annual average for triennium ending 1971	Annual average for triennium ending 1981	Compound annual growth rate, %	Annual average for triennium ending 1971	Annual average for triennium ending 1981	Compound annual growth rate, %
Asia & Pacific	2350	2816	1.8	120.3	127.4	0.6
Rest of the World	2182	2428	1.1	12.5	15.8	2.4
Africa	1830	1766	-0.3	4.0	4.8	1.9
North & Central America	3740	4382	1.6	1.4	2.1	4.2
South America	1667	1841	1.0	5.7	7.3	2.5
Europe	4550	4800	0.5	0.4	0.4	0.0
Others	3190	4633	3.8	1.0	1.2	1.8
World	2331	2767	1.7	132.8	143.2	0.7

Source : FAO 1983

Table 4. Yield and area of rice paddy in countries of Asia and the Pacific Region, 1971-81

Country	Yield (kg/ha)			Area harvested (thousand ha)		
	Ann. av. for trienn. ending 1971	Ann. av. for trienn. ending 1981	Compound ann. growth rate for 1971-81, %	Ann. av. for trienn. ending 1971	Ann. av. for trienn. ending 1981	Compound ann. growth rate for 1971-81, %
Developing Countries						
Bangladesh	1681	1976	1.6	9842	10190	0.4
Bhutan	1098	1103	0.0	38	47	2.1
Burma	1708	2538	4.0	4748	4994	0.5
China	3295	4228	2.5	31341	34320	0.3
Fiji	1839	1768	-0.4	10	9	-1.1
India	1668	1890	1.3	37677	39729	0.5
Indonesia	2346	3317	3.5	8158	8942	1.0
Kampuchea Dem.	1454	932	-4.6	2074	1084	-6.7
Korea DPR	5371	6093	1.3	445	793	5.9
Korea Rep.	4628	5512	1.8	1204	1223	0.1
Laos	1307	1420	0.8	665	720	0.4
Malaysia	2396	2833	1.7	708	755	0.7
Nepal	1937	1828	-0.5	1185	1263	0.8
Pakistan	2246	2450	0.9	1527	1986	2.9
Philippines	1655	2196	2.9	3157	3501	1.0
Sri Lanka	2526	2514	-0.1	579	805	3.4
Thailand	1947	1933	-0.1	6919	8979	2.6
Vietnam	2018	2102	0.4	4916	5547	1.2
All Developing Countries	2271	2712	1.8	117193	24887	0.6
Developed Countries						
Australia	7026	6207	-1.2	38	111	10.7
Japan	5485	5587	0.1	2968	2384	-2.3
All Developed Countries	5505	5615	0.2	3006	2495	-2.0
Asia-Pacific Region	2350	2810	1.8	120199	127382	0.6
Rest of the World	2175	2431	1.1	12677	15785	2.2
World	2331	2768	1.7	132876	143167	0.7

Source : FAO 1983

Thailand, where the productivity declined, relatively. In absolute terms, the yield levels in different countries were remarkably uneven. While there are four countries—Australia, Democratic People's Republic of Korea, Republic of Korea, and Japan—where the yield levels during the triennium ending 1981 were as high as 5.5 t/ha or more, there were a number of countries in the region whose yield levels were barely two tons, or even less. In some countries such as Thailand and Sri Lanka, the increases in production from 1971 to 1981 occurred entirely due to increases in the areas under cultivation, while the yield actually slightly declined.

Outlook for the Future

1. *Increasing and stabilizing rice production in irrigated areas* : Developing countries are making considerable investments on bringing more land under irrigation. Most of the rice cultivated in China is irrigated. In India, nearly 2.5 million hectares are brought under irrigation each year. In the Philippines, about 100,000 hectares are added to the irrigated area every year. Therefore, a priority area of research relates to the optimum use of water so that the yield per litre of water can be maximized. In this context, much research has been done on all aspects of water delivery and on-farm management of water. Methods of achieving equitable distribution of water in the head and tail ends of a command area have been developed.

Prior to the introduction of the Dee-gee-woo-gen and Norin dwarfing genes in rice and wheat respectively, lodging was a serious problem under conditions of good soil fertility. This is why varieties of rice and wheat possessing a semi-dwarf and non-lodging plant type became popular with farmers very speedily. The first high-yielding rice variety developed by IRRI by crossing 'Peta', a tall strain from Indonesia, and 'Dee-gee-woo-gen' resulted in the variety 'IR8'. Soon it was found that when the ecology of the rice field is changed by the application of more water and fertilizer, the threat to yield caused by the triple alliance of weeds, pests and pathogens could become serious. Small farmers obviously can invest on inputs only when risks are low. Therefore, high-yielding varieties characterized by stability of performance were needed. IRRI, hence, introduced in its breeding programme the following approaches in order to combine yield potential with multiple resistance to pests and diseases :

* Continuous identification of new genes for resistance to each of the major disease and insects.
* Sequential release of improved germplasm with different major genes to combat new races of diseases and biotypes of insects.

* Pyramiding of two or more genes into the improved rice germplasm using pedigree method or population breeding approach.
* Gene rotation to reduce the chances of development of races or biotypes.
* Development of multi-line varieties.
* Development of several varieties for a region with different genes for resistance.
* Development of varieties with horizontal resistance.
* Wide hybridization combined with disruptive mating to incorporate genes for resistance from wild germplasm.
* Use of cell culture techniques to develop resistance to toxins produced by major fungal and bacterial diseases.

The pedigree of 'IR 36' which is now cultivated in over 10 million hectares in Asia is an index of the power of the breeding strategy designed to bring about a pyramiding of major genes for resistance. 'IR 36' includes in its ancestry 11 different varieties from 6 countries and a wild species, *Oryza nivara*. IRRI breeders have been able to combine in recent years multiple resistance to pests with earliness and good yield potential (Table 5). Earliness enables multiple cropping wherever there is irrigation.

To make broad-based breeding work possible, there is need for the collection, conservation and evaluation of rice genetic resources. IRRI has, therefore, established an International Rice Germplasm Centre. At present nearly 70,000 strains are available in this collection. This is estimated to represent over 50 per cent of the world's genetic variability in rice. A 5-year plan (1983-87) has recently been developed jointly by IRRI and national research systems to collect the samples of the remaining variability in different parts of the world.

All the collections are systematically screened for nearly 40 different characters under a Genetic Evaluation and Utilization (GEU) Programme.

A great advantage of regional and international collaborative research is the possibility of identifying breeding material suited to different ecological conditions. Under the International Rice Testing Programme (IRTP), the best available varieties from all rice-growing countries are tested in common trials at many locations. Consequently, varieties developed in one country have been found to be suitable for release in some others (Table 6). International cooperation also helps to screen segregating material at "hot spot" locations for major pests and diseases as well as adverse soil factors. Thus, 'IR 36' is the product of selection at a hot spot location for tungro virus in Indonesia, for gall midge in India, and for other pests in the Philippines.

Table 5. Disease and insect reactions of IR varieties in the Philippines

Varieties	Growth duration (days)	Blast	Bacterial blight	Grassy stunt	Tungro	BPH biotypes			GLH	Stem borer	Gall midge
						1	2	3			
'IR 5'	140	MR	S	S	S	S	S	S	MR	MS	S
'IR 8'	130	S	S	S	S	S	S	S	MR	S	S
'IR 20'	125	MR	R	S	MR	S	S	S	MR	MR	S
'IR 22'	125	S	R	S	S	S	S	S	S	S	S
'IR 24'	120	S	S	S	S	S	S	S	MR	S	S
'IR 26'	130	MR	R	S	MR	R	S	R	MR	MR	S
'IR 28'	105	R	R	R	R	R	S	R	R	S	S
'IR 29'	115	R	R	R	R	R	S	R	R	S	S
'IR 30'	110	MS	R	R	R	R	S	R	R	MR	S
'IR 32'	140	MR	R	R	R	R	S	R	MR	MR	S
'IR 34'	130	R	R	R	R	R	S	R	R	MR	S
'IR 36'	110	R	R	R	R	R	R	S	MR	MR	R
'IR 38'	125	R	R	R	R	R	R	MR	R	MR	R
'IR 40'	120	R	R	R	R	R	MR	S	MR	MR	R
'IR 42'	135	R	R	R	R	R	R	S	MR	MR	R
'IR 44'	130	MR	R	S	R	R	R	MR	MR	MR	S
'IR 46'	130	R	R	S	R	R	S	R	MR	S	S
'IR 48'	140	MR	R	R	R	R	R	S	MR	S	S
'IR 50'	105	MS	R	R	R	R	R	MR	R	MR	S
'IR 52'	115	MR	R	R	R	R	R	MR	R	MR	—
'IR 54'	120	R	R	R	R	R	R	S	R	MR	—
'IR 56'	110	R	R	R	R	R	R	R	R	MR	—
'IR 58'	100	R	R	R	R	R	R	R	R	MR	—

R : Resistant; MR : Moderately resistant and S : Susceptible.

Table 6. Some examples of IRTP entries released in different countries

IRTP entry	Country of origin	Country where released
'IET 1444'	India	Nepal
'IET 2935'		Nepal
'IET 2855'		Mali
'Pelita I-1'	Indonesia	Burma, Vietnam
'KN 117'		Burma
'Kn-1b-361-1-8-6-10'		Philippines
'BG 90-2'	Sri Lanka	Burma, Nepal
'BKN 6986-108-3'	Thailand	Burma
'Biplab'	Bangladesh	Vietnam
'BR 51-46-C1'		India
'C 22'	Philippines	Burma
IR 1529-680-3-2'	IRRI (Philippines)	Burma, Mali, Niger, Upper Volta
'IR 36'		Indonesia, India, Philippines
'IR 42'		Cameroon, Nigeria, Philippines
'IR 2307-247-2-2-3'		Vietnam
'IR 2053-57-3-1'		Bangladesh
'IR 2793-80-1'		Bangladesh
'Cica 8'	CIAT (Colombia)	Belize, Guatemala, Honduras, Panama, Paraguay
'Cica 4/IR 665/Tetep'		Sierra Leone

After the release of high-yielding varieties of rice, the production and productivity of rice have gone up considerably in several countries of Southeast Asia. For example, in the Philippines, rice production has been more than doubled between 1960 and 1980. What is even more significant is the doubling of the yield per hectare. The progress made by Indonesia is another outstanding example of the role modern technology can play in improving production and yield, provided the technology package is supported by appropriate packages of services including credit and public policies particularly in the area of pricing and marketing.

The new varieties of rice and wheat are insensitive to photoperiod; hence, they can be cultivated at different times of the year provided water is available and temperatures are favourable. Hence, plant breeders have introduced productivity per day as an important criterion in selection programmes. New crop rotations have consequently become possible. Multiple cropping involving different crop combinations chosen on the basis of their yield potential as well as on marketing opportunities are now being introduced. Rice-wheat rotation has become popular in several parts of India and Bangladesh. Consequently, wheat production has gone up very considerably in Bangladesh and in the West Bengal State of India.

2. *Increasing production under unfavourable environmental condition* : Large areas under rice are either drought prone or flood prone. In recent years, considerable emphasis has been placed on breeding varieties for rainfed upland conditions. The growing conditions in such areas can be classified into 4 broad categories depending upon the number of months when rainfall is received, the moisture-holding capacity of the soil and other soil factors such as alkalinity and salinity.

The technology for stabilizing production under adverse weather conditions include :

* Crop life-saving techniques.
* Alternative cropping strategies to suit different weather models.
* Compensatory production programmes in irrigated areas.

In flood-prone and deep-water (50 cm and above of standing water) areas, there is a need for varieties which can stand submergence and which can divert a larger proportion of the dry matter to grains. At present, the harvest index (i.e., the proportion of total dry matter used for producing grains) is very low in deep-water rices (Table 7). Hence, a better partitioning of the dry matter will have to be achieved in order to increase the grain yield potential. Also under such conditions, it is possible to combine rice cultivation with aquaculture. This will call for the cultivation of varieties with a high degree of pest resistance so that the application of chemical pesticides with long residual toxicity can be avoided.

Table 7. Comparison of grain yield, total dry-matter production and harvest index between irrigated and deep-water rices

Water regime	Variety	Total dry matter (t/ha)	Harvest index	Grain yield (t/ha)
Irrigated lowland	'IR 36' (dry season)	13.7	0.45	7.13
	'IR 36' (wet season)	11.2	0.44	5.74
Deep water	'Kwian Hak'	13.3	0.25	3.29
(floating rice)	'Pin Gaew 56'	12.2	0.19	2.31
	'Hing Hoy'	23.2	0.10	2.23

3. *Problem soil areas* : Over 86 million hectares in South and Southest Asia suffer from various soil health problems. In addition to problems of salinity and other adverse factors, nearly 70 per cent of the rice lands in this region suffer from nitrogen deficiency. Unfortunately, however, there is considerable loss of the applied fertilizer during the monsoon season. Formerly, it was believed that much of this loss is due to leaching and denitrification. However, recent experiments carried out jointly by IRRI and the International Fertilizer Development Centre (IFDC) have shown that a considerable proportion of the loss may be due to ammonia volatilization. Through appropriate placement procedures, the volatilization losses can be greatly minimized and even eliminated. For doing this on a large scale, suitable fertilizer placement equipment will be needed.

A large number of fermers in Asia do not use or use low levels of nitrogen fertilizers because it is either not available or is too expensive. In such a situation, the question arises: how is soil fertility in wetland paddy soils maintained over the years even though a substantial amount of nitrogen is removed by the rice crop, or is lost by leaching, denitrification, and ammonia volatilization? Long-term fertility trials with rice in Japan and in the Philippines suggest that as much as 50 to 75 kg N/ha is added every year to wetland rice fields (Greenland and Watanabe, 1982). Some of this nitrogen comes from exogenous sources like rainfall, irrigation water, and atmospheric ammonia, but biological nitrogen fixation is recognized to be the major source.

Biological nitrogen fixation refers to the property of a few microorganisms to utilize nitrogen from the air, to incorporate it in their cells and to release it in the soil as organic nitrogen through excretion or decomposition after dying.

Compared with dryland soils, wetland rice soils have a high N_2-fixing potentiality. This is because the wetland rice field is a complex ecosystem comprising water and soil, photic and non-photic zones, and oxidized and reduced environments. This diversity permits all the major groups of N_2-fixing organisms to grow in wetland rice fields. Nitrogen-fixing micro-organisms in rice fields comprise :

* Heterotrophic becteria associated with the rice plant and free-living heterotrophic and phototrophic bacteria in the soil.
* Free-living phototrophic blue-green algae.
* Symbiotic blue-green alga *Anabaena azollae*, associated with *Azolla*.

Experiments conducted at IRRI, employing the acetylene-reducion technique, labelled nitrogen, and Kjeldahl technique for nitrogen balance, suggested that a significant amount of dinitrogen is fixed in rice fields by these N_2-fixing microorganisms and that a part of it is rapidly utilized by the rice plant. Research carried out at different centres in India has revealed considerable possibilities for using bio-fertilizers in rice production.

It is thus possible to harness various organic and bio-fertilizers such as *Azolla*, blue-green algae, straw, waste materials and legume nitrogen. *Azolla* is already used fairly extensively in China and Vietnam. Experiments carried out at IRRI have shown that if the following three conditions are followed, *Azolla* application would save to some extent expense on mineral fertilizer:

* Adequate phosphorus availability in the soil.
* Effective control of *Azolla* pests, whenever necessary.
* Favorable labour wage rate.

The scope for introducing nitrogen-fixing genes in rice through genetic engineering techniques is being explored. However, this pathway of minimizing the need for mineral fertilizers is not likely to become available for commercial use during this decade.

4. *Adding a nutritional dimension to cropping systems research* : A major scientific challenge is to make the leguminous plant utilize more energy. Many grain legumes which are energy-rich are unfortunately grown under conditions of energy deprivation. Success has been achieved in reducing the duration of several grain legumes like cowpea and pigeonpea. Research on improving the yield potential of grain legumes needs intensification. Legumes can find a place in many inter-cropping and multiple-cropping sequences having rice as the major crop.

Emerging Scenario

Many experts believe that by 2015 A.D., the global population will reach about 8 billion. A considerable part of this increase will take place in the rice-growing and rice-consuming nations of Asia. Can we continue to improve the yield of rice and produce more and more from less and less land ? In its study "*Agriculture in 2000*", FAO has calculated that the total rice requirements of the world will be of the order indicated in Table 8. Since the gap between potential and actual yields even at currently available levels of technology is wide in farmers' fields, it should be possible to produce the quantities indicated in Table 8 through the spread of modern technology in the fields of small farmers. This will call for carefully designed packages of services and public policies.

Scientists are examining various possibilities of raising the ceiling to yield further. Hybrid rice technology has attracted much attention following the development and release of hybrid rice varieties in China, where about 6 million hectares are currently planted to hybrid rice. It has been reported that hybrid rices have yielded 20-30 per cent more than the best semidwarf varieties under irrigated conditions. Encouraged by these developments in China, IRRI intensified

Table 8. Projected demand for rice (unmilled) in the year 2000 A.D. with zero, high, and medium rates of income growth and specified income elasticities

Country	1980 demand[a]	Projected demand (thousand t) with income growth rate[b]			% change with medium
		Zero	High	Medium	
China	138,007	183,367	212,432	206,478	50
India	79,552	113,430	129,451	125,147	57
Indonesia	33,105	43,430	68,495	59,488	80
Bangladesh	21,445	32,513	39,488	36,988	72
Thailand	14,119	20,632	21,487	22,341	58
Burma	9,166	13,353	15,228	14,578	59
Philippines	7,042	10,688	12,509	12,387	76
Sri Lanka	3,033	3,429	4,119	4,009	32
Total of above	305,389	417,842	503,099	481,526	58
Total of Asia	360,765	493,609	594,326	570,008	58

[a]Production + imports – exports. [b]Income elasticities

Source : Herdt and Stangel 1982

research on hybrid rice in 1979 with the objective of exploring the prospects and problems of using this technology for increasing yield potential of this crop in the tropics. Subsequently, the national rice improvement programmes in Indonesia, India and South Korea have also joined IRRI to collaborate in hybrid rice research.

The results obtained to date indicate that rice hybrids provided significant yield advantage (14 to 34 per cent) compared with the best check varieties. The yield advantage in the hybrids was attributable to increase in dry-matter production (source size) and/or higher harvest index, due to increased number of spikelets per unit area and 1000-grain weight (sink size). The hybrids were also superior to check varieties for productivity per day and thus appear to be more efficient physiologically. They also provide opportunity for expeditious pyramiding of dominant genes for resistance to biological stresses. The better-developed root system and increased vegetative vigour of the F_1-hybrids should also make them adaptable to rainfed lowland conditions. This needs to be explored, however. Genetic tools, viz., cytoplasmic male sterile (CMS), maintainer and restorer lines, essential to develop F_1-hybrids in this strictly self-pollinating crop are available and can be adapted to tropical conditions. Frequency of maintainer and restorer lines is high (10 and 25 per cent, respectively) among the elite lines developed through conventional breeding methods in the tropics. These lines can be used as parental materials to develop the desired combinations. Therefore, hybrid breeding would supplement the conventional breeding efforts to breed rice varieties with higher yield potential.

The major bottleneck in hybrid rice technology is the seed production. The farmers have to buy hybrid seed every crop season and its cost is going to be higher than of the non-hybrid varieties. The seed production techniques, developed in China, although found adaptable under experimental conditions in the tropics, have yet to be tested under large-scale field conditions. Being labour-intensive and skill-oriented, the adoption of these techniques would depend on the availability and cost of labour. Under the Asian situation it may not be a serious problem.

The technological issues which require resolution during the current decade include:

* Identification of ecological and cultural conditions in which rice hybrids would manifest significant yield superiority over conventionally bred varieties.
* Level of disease and insect resistance of F_1-hybrids compared with non-hybrid varieties.
* Development of new genetic tools to breed productive hybrids.
* Identification of countries where hybrid seed production techniques can be adopted, and organizations can be developed to produce, certify, and market the seeds.

* Finding out ways through which seed yields in hybrid seed production plots can be increased.
* Agronomic management of hybrids compared with non-hybrids.
* Economics of hybrid seed production.
* Study of possible socio-economic consequences of hybrid rice technology.

Hopefully, the scientists involved in hybrid rice research would find answers to these issues during the current decade. Results obtained by them would determine how far hybrid breeding technology would be adopted in major rice-growing countries in the world by the year 2000 A. D.

Some of the the other developments which we may witness towards the end of the century are the following:

* *Oryza glaberrima* may disappear as a cultivated species and will be replaced by *O. sativa*. However, some useful traits from *O. glaberrima* will be incorporated into varieties of *O. sativa*.
* Within the *O. sativa* complex, *javanica* varieties may no longer be grown by the year 2000. There is no work going on for the improvement of *javanica* rices and these rices are being rapidly replaced by *indicas*. Many important traits will of course be incorporated into *indicas*.
* The available variability in *glaberrima* and *javanica* rices will be preserved in the International Rice Germplasm Centre of IRRI.
* Area planted to *japonica* rices will decrease as evidenced by replacement of *japonica* varieties by improved *indicas* in Korea and China.
* Short-duration varieties, of 100-110 days growth duration, with multiple disease and insect resistance will be planted in much of the irrigated areas, thereby promoting multiple cropping.
* There will be tremendous pressure to bring rainfed areas under irrigation especially in the food deficit countries. Similarly in some of the deep-water areas new cropping patterns will emerge whereby there will be no need to plant rice during the flood-prone season. Considerable land formerly planted to deepwater rice in Bangladesh is now left fallow during rainy season but a crop of boro rice is planted instead. These developments will lead to drastic reduction in the number of photoperiod-sensitive varieties.
* Varieties of the future will have much better grain quality characterized by intermediate amylose content and intermediate gelatinization temperature. Some with aromatic grains will be grown. The improved palatability will also lead to greater per capita consumption.

* Chemicals will be used for increasing protein content beyond the genetic potential and with fertilizer management. If genetic potential of rice is to produce 8 per cent protein, then additional 2 per cent point protein can be increased by the use of nitrogen fertilizer at optimum rate and timing, which will make the protein content to 10 per cent. Several studies suggest that the use of some herbicides such as triazines and substituted ureas if applied at sub-herbicidal level can increase protein content of rice grain by additional 1 per cent point which will make a total of 11 per cent.
* Growth regulators in rice are likely to be used in early 2000 A.D. to increase both quantity and quality of tropical rice production.
* Use of urease inhibitors such as PPD or similar compounds may be used to modify nitrogen transformation processes and therefore increase nitrogen use efficiency and grain yield of rice.
* Following the introduction of extremely short-duration rices and controlled irrigation water, farmers may continue to change from transplanting to direct seeding.

Apart from the advances in improving yield and quality, another area which is likely to undergo great changes is the whole field of post-harvest technology. Advances in chemical engineering and technology have helped in standardizing methods which could provide value-added products from every part of a plant or animal. I would like to illustrate this from the possibilities now open in the utilization of rice biomass.

1. *Rice straw*—as a source of manure for soil incorporation, a substrate for mushroom culture, a source of balanced animal nutrition for stall-fed animals (rice straw can be subjected to alkali to digestion and enriched with urea and molasses), a source for the generation of biogas and composting, and as a cellulosic raw material for paper manufacture.

2. *Rice bran*—extraction of edible rice bran oil for human use, use of bran oil in the manufacture of soaps and other products, use of deoiled rice bran for poultry feed, and fortification of deoiled rice bran to improve animal nutrition.

3. *Rice hull or husk*—as energy for grain drying and charcoal production, for animal feed after appropriate fortification, use of rice hull ash for soil incorporation and hollow building blocks, and as a source of solar grade silicon for use in the manufacture of photovoltaic systems.

Similar analysis can be made in the case of all crops and animals. An optimum blend of production and post-harvest technologies could help improve rural incomes and employment.

Thus, the overall outlook is one of hope and optimism. There is, however, no. time to relax since new problems of plant health and soil care arise all the time. Only a dynamic research programme can sustain a dynamic production programme. Fortunately, India is blessed with such a dynamic research system in rice through the All-India Co-ordinated Rice Improvement Project of the Indian Council of Agricultural Research. The growth of such a dynamic rice research programme owes much to the vision and work of great scientists like Dr. K. Ramiah and Dr. N. Parthasarathi. To Dr. B. P. Pal, we owe a deep sense of gratitude for his leadership in the organisation of All-India Co-ordinated Research Projects.

References

CHANG, T.T. 1976. The origin, evolution, cultivation, dissemination and diversification of Asian and African rices. *Euphytica* 25, pp 425-441. Food and Agriculture Organization 1983. *Cereal Production in Asia and the Pacific for the Period 1971-81.* FAO, Bangkok, Thailand, pp. 21.

Greenland, D.J. and Watanabe, I. 1982. The continuing nitrogen enigma. In *Wither Soil Research*, Trans 12th Int Congr Soil Sci. in New Delhi, pp. 123-137.

HERDT, R.W. AND CAPULE, C. 1983. *Adoption, Spread and Production Impact of Modern Rice Varieties in Asia.* IRRI, Los Banos, Philippines, pp. 54.

HERDT, R.W. AND STANGEL, P.J. 1982. Population, rice production and fertilizer outlook. Paper presented at the Workshop on Organic Matter and Rice, 28 September-October 1982. IRRI, Los Banos, Philippines.

PORTERES, R. 1956. Taxonomie agrobotanique des rix cultives *O. sativa* Linne. et *O. glaberrima* Studel. *J. agric. Prop. bot. Appl.* **3**: 341-384, 541-580, 627-700, 821-856.

OKA, H.I. AND MORISHIMA, H. 1971. The dynamics of plant domestication: cultivation experiments with *Oryza perennis* and its hybrid with *O. sativa*. *Evolution* **25**: 356-364.

Future Thrusts

In most of the regions, climate variability appears to have significant impact on rice production. During the recent past, there has been advancement of knowledge in understanding and predicting the important facets of variability of climate as well as understanding and modeling of crops and farming systems. The time is thus opportune to harness the advances in the relevant disciplines for management for sustained high production in the face of climate variability.

For identifying management strategies for maximum production over a specific agro-climatic zone, it is necessary to adopt a multi-pronged approach involving (i) analysis of the observed variation in production and yield on spatial scales ranging from agricultural field station and farmers field to district, state and beyond (ii) detailed studies of the relationships between production/yield, abiotic stresses and climate. (iii) techno-management options available to the farmers and (iv) developments of models for the total impact of climate on production.

Most of the rice models do not take into consideration moisture stress experienced by rainfed rice. Studies on the inter-comparison of the performance of the major rice models in simulating the observed inter-annual variation in the production at the field stations in China, Japan and Philippines showed that skill at simulation was rather poor even for known conditions at the field station. Models which include all components of the impact of climate adequately including the impact on triggering of pests and diseases needs to be developed so that they become capable of generating realistic simulation of the inter-annual variation in productivity. For this, basic information on the biology of causal organisms, weather influence on host and pathogen or insect pest and on the modern crop production practice is available. Summary models that capture the essential components of system dynamics held the key to success in forecasting a disease or insect outbreak in India.

Climate change due to enhanced concentration of green house goses is likely to have an impact on rice production. Genetic variability for tolerance to high temperature and response to elevated CO_2 suggest that germ plasm improvement is possible for meeting the challenge of climate change by improving adaptation of crops to changed climate.

On the other hand, rice production will have impact on climate because of methane concentrations. There is a large uncertainty in the estimation of sources and sinks of methane. In depth studies on basic aspects that influence production, consumption and transport particularly in relation to developing mitigation strategies and realistic estimates of contribution of rice culture to anthropogenic methane emission are required.

Editors

List of Contributors

1. **NM Nayar**
Central Tuber Crops Research Institute
Thiruvananthapuram 695 017

2. **S Gadgil, K Narahari & S Pradeep**
Centre for Atmospheric and Oceanic Sciences
Indian Institute of Science
Bangalore 560 012

3&4. **BC Biswas, HP Das & SK Saha**
Agrometeorology Division
Indian Meteorology Department
Pune 411 005

5. **ASRAS Sastri, SK Rai, D Naidu & AK Srivastava**
Indira Gandhi Agricultural University
Raipur 492 012

6. **N Singh, A Sontakke & SK Patwardhan**
Indian Institute of Tropical Meteorology
Pune 411 008

7. **EA Siddiq**
Directorate of Rice Research
Rajendranagar
Hyderabad 500 030

8. **V Balasubramanian**
Directorate of Rice Research
Rajendranagar
Hyderabad 500 030

VP Singh, A Sharma, NVK Chakravarty & YP Abrol
Division of Plant Physiology and Agricultural Physics
Indian Agricultural Research Institute
New Delhi 110 012

9. **PK Aggarwal**
Division of Environmental Sciences
Indian Agricultural Research Institute
New Delhi 110 012

AK Singh
Water Technology Centre
Indian Agricultural Research Institute
New Delhi 110 012

N Kalra
Division of Environmental Sciences
Indian Agricultural Research Institute
New Delhi 110 012

TM Thyagrajan
Tamil Nadu Agricultural University
Coimbatore 641 003

PR Reddy
Central Rice Research Institute
Cuttack 753 006

S Mohandas
Tamil Nadu Rice Research Institute
Aduthurai 612 101

10. **IC Pasalu, NV Krishanaiah & K Muralidharan**
Directorate of Rice Research
Rajendranagar
Hyderabad 500 030

CD Mayee
Marathwada Agricultural University
Aurangabad 431 005

K Muralidharan, S Venkatraman, NVL Rajarajeshwari
Directorate of Rice Research
Rajendranagar
Hyderabad 500 030

K Manibhushan Rao & S Natarajan
Centre for Advanced Studies
University of Madras
Chennai 600 025

Jayant Sarkar
Agrometerology Division
Indian Meteorological Department
Pune 411 005

11. **DC Parashar**
Chemistry Division
National Physical Laboratory
New Delhi 110 012

N Sethunathan, S Mishra, TK Adhya, B Ramakrishnan, VR Rao
Central Rice Research Institute
Cuttack 753 006

S Anthoni Raj & JB Ranganathan
Agricultural College
Madurai

S Thayumanavan
Tamil Nadu Rice Research Institute
Aduthurai 612 101

12. **MS Swaminathan**
International Rice Research Institute
Los Banos, Philippines
Present address: MS Swaminathan Research Foundation
Chennai